Distillation Topical (Including Papers from the Separations Division)

Topical Conference at the 2009 AIChE Spring National Meeting

Tampa, Florida
26 – 30 April 2009

ISBN: 978-1-61567-260-8

Printed from e-media with permission by:

Curran Associates, Inc.
57 Morehouse Lane
Red Hook, NY 12571

Some format issues inherent in the e-media version may also appear in this print version.

Printed by Curran Associates, Inc. (2009)

For permission requests, please contact AIChE
at the address below.

AIChE
3 Park Avenue
New York, NY 10016-5991

Phone: (203) 702-7660
Fax: (203) 775-5177

www.aiche.org

Additional copies of this publication are available from:

Curran Associates, Inc.
57 Morehouse Lane
Red Hook, NY 12571 USA
Phone: 845-758-0400
Fax: 845-758-2634
Email: curran@proceedings.com
Web: www.proceedings.com

TABLE OF CONTENTS

DISTILLATION PAPERS

SEPARATION PAPERS

15a How to Properly Evaluate and Document Tower Performance

Daniel R. Summers, Sulzer, 4019 South Jackson Avenue, Tulsa, OK 74107

Obtaining good tower performance information is a difficult and tedious task that needs special attention to many details. The end result of such an endeavor can lead to a wealth of knowledge that can be invaluable for the operation of that particular tower and numerous others. How to generate a good performance evaluation with proper sensitivity analysis and documentation will be the topic of this paper. Special attention will be given towards model choice, Heat and Material balancing, data acquisition, data evaluation, hydraulics evaluation, performance diagram generation, sensitivity analysis and final documentation.

The author has had the opportunity to evaluate numerous towers to examine both good and bad operation. Many of these evaluations have been the topics of his past AIChE papers. The wealth of his experience will be placed in this one document to help establish a standard for future tower evaluations.

Simulation and Optimization of Depropanizer, C3-Splitter and Debutanizer

Bora **Aydin** and Marzouk **Benali**[*]

Natural Resources Canada, CanmetENERGY, 1615 Lionel-Boulet Blvd., P.O. 4800, Varennes, QC, J3X 1S6, Canada

Abstract

In this study an energy analysis of depropanizer, C3-splitter and debutanizer is presented for an industrial ethylene production plant with a yearly ethylene throughput of 500,000 tons. Simulation and optimization was done based on the industrial data by using CHEMCAD software. According to the observed effects of thermodynamic and K-value model combinations on the desired product yields, optimum model combination was selected. The reboiler and condenser duties of depropanizer were optimized to separate C3- and C4-products completely. Optimum values of reboiler and condenser duties were explored to obtain maximum purity and flow rate for the desired components in the overhead of C3-splitter and debutanizer.

Keywords: Ethylene, optimization, energy analysis, depropanizer, C3-splitter, debutanizer

Background

The ethylene production system consists of highly complex sub-systems. The simulation and/or optimization of the whole system could generate severe problems. It is imperative to divide the process into the sub-systems for efficacious optimization purposes. On the basis of a sensitivity analysis, the most sensitive thermodynamic model parameters from each critical separation task are tuned in accordance with the industrial data and the product separation requirements. The most difficult, expensive, and energy-intensive

[*] Corresponding author.
Email: mbenali@nrcan.gc.ca ; Tel.: (450) 652-5533; Fax: (450) 652-5918

separations are demethanizer, depropanizer, debutanizer as well as ethylene/ethane (C2) and propylene/propane (C3) splitters. In open literature simulation efforts and their significant impact on control mode was emphasised for energy optimization purposes (*e.g.* Gonzales and Ferrer, 2006; Yamashita, 2000; Kesler and Friedman, 1998).

The separation of multicomponent mixtures by distillation can be difficult when the Vapour-Liquid Equilibrium (VLE) system is non-ideal and azeotropes are present. The products obtained from various distillation towers could be estimated via simulation of the separation systems. The purity of components and the composition of the output depend on the thermodynamic models. Thermodynamic modeling and simulation efforts were performed for an industrial ethylene production plant with a yearly ethylene throughput of 500,000 tons. Depending on the composition of the feedstock and the operating conditions (feedstock temperature, column top pressure, column pressure drop, number of stages, feed stage position) different K-Value and enthalpy model combinations were suggested for each distillation unit. Optimization tool was implemented on selected distillation towers (*i.e.* depropanizer, C3-splitter, debutanizer) to increase the propane (C_3H_8), propylene (C_3H_6), 1,3-butadiene (C_4H_6), 1-butene (C_4H_8), i-butene (i-C_4H_8), n-butane (n-C_4H_{10}) yields as well as to decrease the condenser and/or reboiler duties. Energy consumption analysis resulted with the optimization of depropanizer, C3-splitter and debutanizer.

The feed to these fractionation trains is composed of ethane, propane, acetylene, ethylene, propylene, n-butane, i-butene, 1-butene, 1,3-butadiene. In this work, the effect of SRK, PSRK, and Peng-Robinson (PR) models on the separation performance of the depropanizer was studied.

C3 and C4 Separation Analysis

In the ethylene plant considered the products to be separated at consecutive fractionation are listed as hydrogen and methane fractions, high-purity ethylene and propylene, ethane and propane fractions, C4-products and gasoline.

The dried cracked gas is chilled and sent to the demethanizer and the cold-box sub-system (*i.e.* the cascade refrigeration system). The demethanizer bottoms are directed to the deethanizer where C2-products are separated and forwarded to the overhead. The deethanizer overhead is hydrogenated in a 5 m^3 catalytic reactor to remove acetylene. Purified hydrogen is added to C2-products entering the reactor. The ethylene-ethane stream is carried from the acetylene converter to C2-splitter. The deethanizer bottoms are transferred to depropanizer where C3 components are separated from the heavier C4-products. The overhead from this fractionator is fed to a two-stage catalytic reactor (10 and 40 m^3 in volume respectively) to convert methyl acetylene and propadiene to propylene. Since polymer grade propylene is required, the C3 fraction from this reactor flows to C3-splitter to separate propylene from propane. Reflux of high pressure towers is condensed with cooling water, whereas low pressure towers require refrigerant-cooled condensers. The net overhead from the column is high-purity propylene and the bottoms is mainly propane, which is recycled to the cracking furnace. The bottoms from the depropanizer is transferred to the debutanizer, where the C4 fraction, containing C_4H_6, C_4H_8, i-C_4H_8, and n-C_4H_{10}, is separated as an overhead from the C5 and heavier components. Reflux is condensed with cooling water, and the reboiler is indirectly heated with the low pressure steam.

Simulation and Optimization Methodology

The simulation and optimization of the three distillation towers were performed by using CHEMCAD software. For simulation purposes, CHEMCAD offers two different column configurations: tower, and simultaneous correction distillation (SCDS) columns. After preliminary simulation test results based on the industrial data, tower was chosen as depropanizer and SCDS was chosen as C3-splitter and debutanizer. Stoichiometric reactor model was used for the simulation of methyl acetylene and propadiene hydrogenation. At constant operating conditions (temperature, pressure, column pressure drop and industrial feed composition) the separation units were simulated to acquire optimal feed position and reboiler and condenser heat duties for the energy analysis.

Thermodynamic and Energy Analysis of Depropanizer

Table 1 shows the effect of enthalpy models on the overhead and bottoms compositions of the depropanizer. For a given K-Value model the top and bottom compositions obtained by the simulation of a distillation unit depend on the chosen thermodynamic model. Similar sensitivity analysis was performed with the above mentioned K-Value models for different separation units. Tray, distillate and residue compositions were analyzed in detail for suggested thermodynamic models.

Table 1. Effect of enthalpy models on the overhead and bottoms compositions of the depropanizer: SRK as K-Value Model

	SRK	LH	PR
Propane (Top) (kmol/h)	360.27	360.83	368.36
Propane (Bottom) (kmol/h)	8.09	7.53	0.00
Propylene (Top) (kmol/h)	541.57	541.61	541.91
Propylene (Bottom) (kmol/h)	0.34	0.29	0.00
1,3-Butadiene (Top) (kmol/h)	41.69	41.69	17.94
1,3-Butadiene (Bottom) (kmol/h)	40.21	40.19	63.96

According to the simulation results, SRK-PR model combination was selected for the depropanizer to obtain higher mole fraction of C_3H_6 and C_3H_8 in the overhead and higher C_4H_6 mole fraction in the bottoms. Same type of analysis was done for the C3-splitter and the debutanizer and SRK-SRK model combination was selected for both SCDS column configurations.

As seen in Fig.1, the depropanizer was simulated at various reboiler and condenser duty values for optimization purposes to obtain maximum purity for C_3H_6 in the overhead (Fig. 1a) and that for C_3H_8 in the bottoms (Fig. 1b). Appropriate values for the reboiler and condenser duties (2.9 MW and -2.9 MW respectively) were selected based on the deviation of the mole fraction values of C_3H_6 and C_3H_8 between the maximum and the values obtained at 2.9 MW and -2.9 MW.

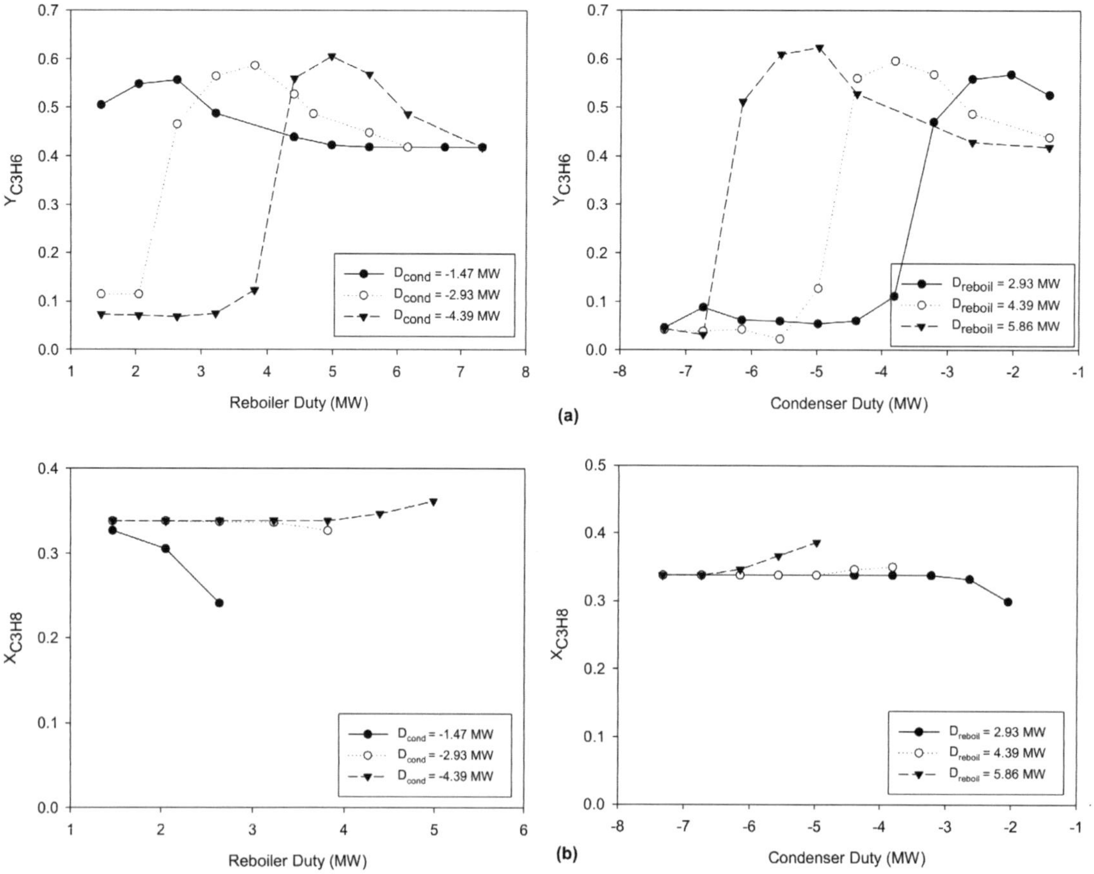

Fig. 1 Mole fraction of propylene in the overhead (a) and of propane in the bottoms (b) for various reboiler and condenser heat duty values.

Optimization of Reboiler and Condenser Heat Duties for Debutanizer

The efficiency of C3-splitter could be specified by means of C_3H_6 mole fraction in the overhead (Y_{C3H6}) and C_3H_8 mole fraction in the bottoms (X_{C3H8}) as well as product throughput. Optimal C3-splitter operation could be defined as the maximum flow rate and purity of C_3H_6 in the overhead and that of C_3H_8 in the bottoms with minimum energy requirements.

Fig. 2a shows the effect of the condenser duty on the mole fractions of C_3H_6 and C_3H_8 in the overhead (Y_{C3H6}, Y_{C3H8} respectively) at a given reboiler duty. As seen in Fig. 2a, at a condenser heat duty of -6.4 MW, the mole fractions of C_3H_6 and C_3H_8 in the overhead are 0.97 and 0.02 respectively. However, the mole fraction of C_3H_6 in the bottoms is 0.61

(Fig. 2b). At a condenser heat duty of -5.3 MW, the mole fraction of C_3H_6 in the bottoms (X_{C3H6}) is zero but the purity in the overhead was found to be below 0.9. Thus, -5.9 MW was selected as the optimum value for the condenser duty to obtain a high C_3H_6 purity in the overhead (Y_{C3H6}=0.92) and low C_3H_6 fraction in the bottoms (X_{C3H6}=0.45). Though Y_{C3H6} is 0.92 additional fractionation step is required; the ratio of the amount of C_3H_6 in the bottoms to that in the overhead is 0.33.

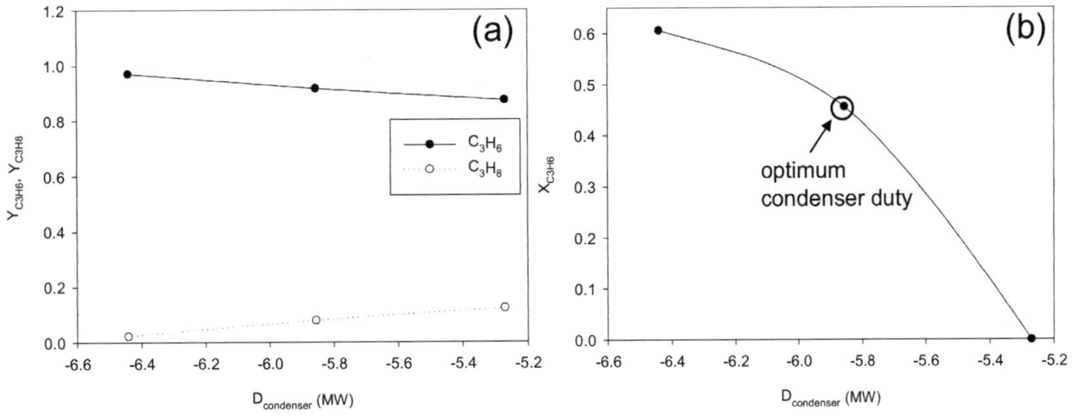

Fig. 2 Influence of the condenser duty on (a) the mole fraction of C_3H_6 and C_3H_8 in the overhead (b) the mole fraction of C_3H_6 in the bottoms. $D_{reboiler}$ = 7.3 MW.

Mole fractions of C_3H_6 and C_3H_8 in the bottoms are plotted in Fig. 3a for various reboiler duties at constant condenser duty. X_{C3H6} decreases while X_{C3H8} increases with increasing reboiler duty. At a reboiler heat duty of 7.9 MW, the mole fractions of C_3H_6 and C_3H_8 in the bottoms reach their minimum and maximum values respectively. At the "point of crossing" the mole fraction of both components in the bottoms is 0.25. At the same reboiler heat duty the corresponding mole fraction of C_3H_6 in the overhead is 0.89. As indicated above the main objective of C3-splitter is to obtain C_3H_6 in the overhead with high purity (*i.e.* 0.99). At the "point of crossing" the mole fraction of C_3H_6 is 10% lower than the desired purity. At 7.3 MW the mole fraction of C_3H_6 in the overhead is 0.92 (Fig. 3b), 7% lower than the targeted value. The mole fraction of C_3H_6 in the bottoms is 0.45, which confirms the need of an additional fractionation step.

Thus for the given feedstock for this study, the optimum condenser and reboiler duties of C3-splitter are -5.9 MW and 7.3 MW respectively.

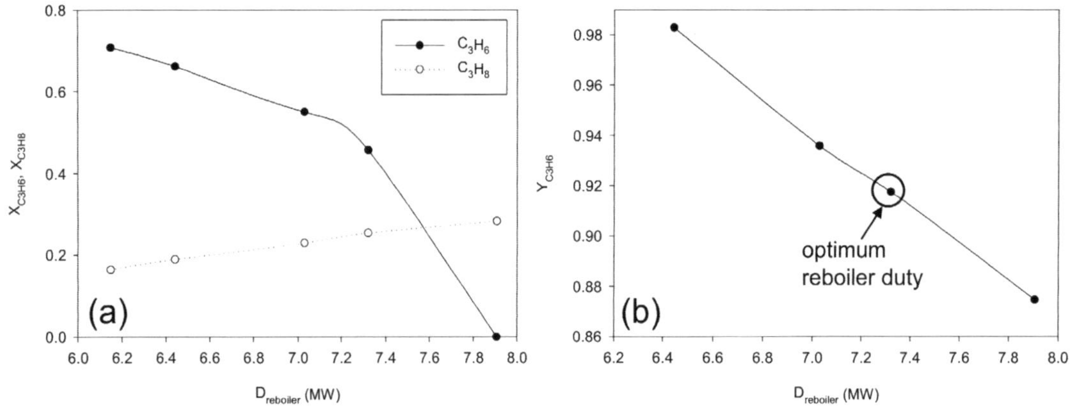

Fig. 3 Effect of the reboiler duty on (a) the mole fraction of C_3H_6 and C_3H_8 in the bottoms and (b) the mole fraction of C_3H_6 in the overhead. $D_{condenser}$ = - 5.9 MW.

As seen from Fig. 1, the feed is sent from the bottoms of the depropanizer to the debutanizer where C4-products are separated and sent to the overhead. The components of the feed are mainly C_4H_6 and significant amount of C_4H_8, i-C_4H_8 and n-C_4H_{10}. Fig. 4 shows the influence of the condenser duty on the component mole fraction in the overhead. The difference in the mole fractions of C_4H_8, i-C_4H_8 and n-C_4H_{10} obtained at -0.8 MW and -0.03 MW is low. In addition, the mole fraction of C_4H_6 in the overhead increases with decreasing condenser duty. Thus -0.03 MW is the required condenser duty for the debutanizer to obtain highest C4 fractions' purity and flow rate in the overhead.

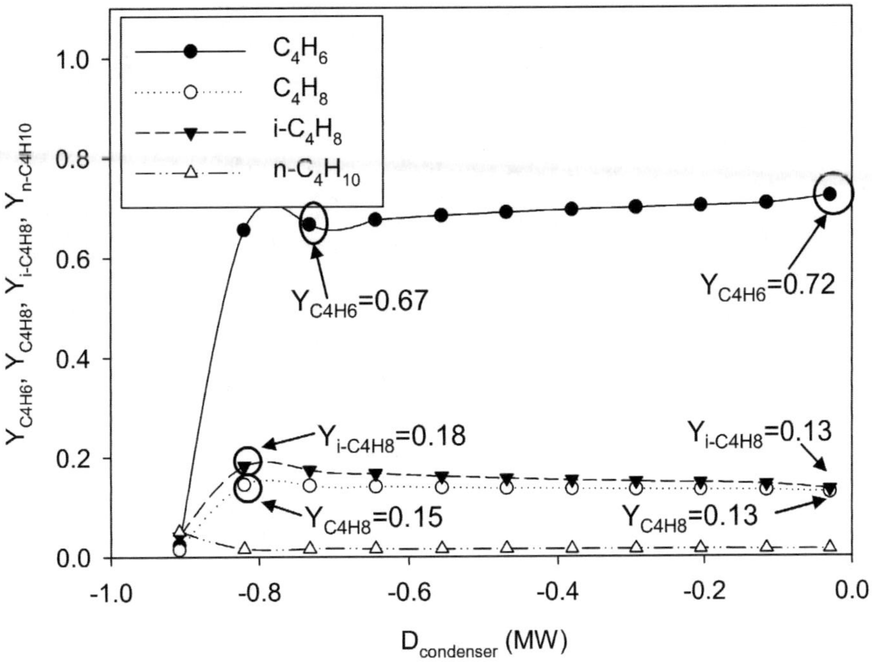

Fig. 4 Composition of the debutanizer overhead as a function of the condenser duty. $D_{reboiler}$ = 0.6 MW.

The effect of reboiler duty on the mole fractions of C_4H_6, C_4H_8, i-C_4H_8 and n-C_4H_{10} is illustrated in Fig. 5a. The mole fractions of all components are constant for all reboiler duties. Thus 0.15 MW could be selected as the required reboiler duty. On the other hand recovery fraction (R), which is an indicator of the separation efficiency (Eq. 1), increases with the reboiler duty (Fig. 5b). The desired optimal operation of a debutanizer could be described as the maximum purity and flow rate of C4 fractions in the overhead. Thus 0.6 MW of reboiler duty is required for the complete separation of C_4H_6, C_4H_8, i-C_4H_8 and n-C_4H_{10} (R=0.99).

$$R_A = 1 - \frac{\dot{m}_{A,bottoms}}{\dot{m}_{A,feed}} \qquad \text{(Eq. 1)}$$

where,

R_A is recovery fraction of component A, $\dot{m}$ is mass flow rate of component A.

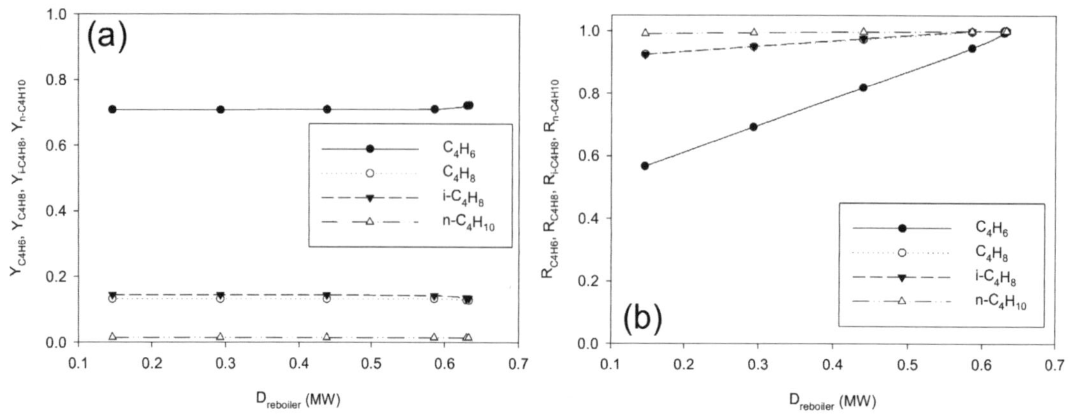

Fig. 5 Influence of the reboiler duty on (a) the mole and (b) the recovery fractions of C_4H_6, C_4H_8, i-C_4H_8 and n-C_4H_{10}. $D_{condenser}$ = - 0.03 MW.

Concluding Remarks

In this study energy analysis was applied to three separation units of the ethylene production plant. For the given K-value model (SRK), the influences of thermodynamic models (SRK, LH, PR) on the separation efficiency were demonstrated for the simulation purposes. SRK-PR is the appropriate combination for the depropanizer whereas SRK-SRK gives the highest product recovery for C3-splitter and debutanizer. The reboiler and condenser heat duties were optimized to obtain the desired product purity and yield specific to each distillation design. The optimum condenser and reboiler duties for maximum C_3H_6 recovery are -5.9 MW and 7.3 MW, respectively. CHEMCAD was proved to be a powerful tool for the simulation and optimization of industrial systems.

References

Gonzalez, R.; Ferrer, J.M., Analyzing the value of first-principles dynamic simulation, *Hydrocarbon Processing*, 9, 69-75, 2006.

Kesler, M.G.; Friedman, Y.Z., Model based control of fractionation in an ethylene plant, *AIChE Spring National Meeting*, 1998.

Yamashita, Y., Supervised learning for the analysis of process operational data, *Comp. Chem. Eng.*, 24, 471-474, 2000.

15c Distillation Trouble... Shot - Some of BP Chemicals' Experiences

Sean Hennigan, Manufacturing Excellence and Technology Implementation Team, BP Chemicals Limited, BP Hull Research and Technology Centre, Saltend, Hull, HU12 8DS, United Kingdom

Some examples of column successes and surprises within BP Chemicals experience. Includes operational troubles that have been diagnosed by applying usually simple methods to see beyond the immediate symptoms

15d An Overview of Distillation Trays and Packings in Japan

Shuzo Ohe, Science University of Tokyo, 1-3 Kagurazaka, Shinjuku-ku, Tokyo, 162-8601, Japan

In Japan, various kinds of trays and packings have also been used for distillation, as in the States. Some of them are commonly used in various. However, there are some trays and packings which are developed and used only in Japan. For example, the Lift Tray, which is famous in Japan, is not known in the States. This tray is widely used in Japan and other Asian countries but has not been tested at the FRI. In the 1970s, large companies became interested in the development of new trays, for example, the TFT, the new VST and the Angle Tray. However, after the oil crisis, these companies no longer developed new trays or packings by them. Around the same time, a well known structured packing appeared which further reduced the motivation of large companies to develop new trays and packings. Instead, small scale companies started to develop new trays and packings. Some of these companies have succeeded commercially and are continuing to be used in Japan.

15e Reduce Energy Demands in Industrial Distillation Trains through the Use of Thermal Links

Josue Santos-Mendez, Instituto Tecnologico Superior de las Choapas, 96980, Las Choapas, veracruz, Las Choapas, Ver Orquidea 109 Col, Mexico and Salvador Hernandez, Chemical Engineering, Universidad de Guanajuato, Noria Alta S/N, Guanajuato, Mexico.

In this paper the retrofit of five conventional distillation trains for the separation of quaternary mixtures of hydrocarbons for feed compositions with high or low content of intermediate components is presented. The retrofit only requires the introduction of liquid or vapour recycle streams between the conventional distillation columns. Each recycle stream removes one condenser or one reboiler. The introduction of thermal links can lower the energy consumption up to 40% in contrast to conventional distillation trains widely used in the industry. This efficiency in the use of energy is achieved because the recycle streams, introduced in the conventional distillation trains, reduce the remixing associated with higher energy consumption.

Keywords: Distillation trains, energy savings, thermal link, remixing

15f Energy Saving by Simply Feed Preheating at Petrochemical Plant

Tadao Matsumoto, Petroleum Refining Technology Center, Idemitsu Kosan Co., Ltd., 2-1, Anesaki-Kaigan, Ichihara City, Chiba, 299-0107, Japan

In general, the feed fluid of distillation column is preheated by utilizing feed/bottom heat exchanger or heat recovery system and the feed condition is close to bubble point. The reboiler load could be reduced by installation of a new preheater which uses cheap low-pressure steam. In order to maintain separation of the column, reflux flow rate should be a little larger than before installing the preheater. Therefore, reboiler duty reduction (ΔHR) becomes smaller than feed enthalpy increase(ΔHF) and this ratio, ΔHR/ ΔHF depends on application systems.

Two practical energy saving jobs relating to feed conditions were performed at Chiba facility of Idemitsu Kosan Co. in Japan. First one is a benzene column of aromatics plant. A new feed preheter was installed and made the feed approximately 90 percent vaporization to reduce the reboiler whose hot medium was hot-oil.

Second job was carried out in styrene monomer plant. When ethyl-benzene production process was converted from aluminum chloride process to liquid phase zeolite process, reboiler duty of the two distillation columns could be reduced by eliminating water washing section.

We investigated necessity of modifications including column internals and auxiliary using process simulator during the basic planning phase of these jobs and finally decided to modify feed pipes and feed nozzles for these three columns. These distillation columns are running well as we expected.

26a Corrosion in Crudes, Myths and Reality

German A. Luna Sr., Process Department, FLUOR CANADA, 55 SUNPARK PLAZA SE, Calgary, AB T2X3R4, Canada

Corrosion problem is a very common and extensive conversation topic between process engineers and operators. HCl corrosion by hydrolysis of Ca, Mg salts is the most known form of corrosion; use of caustic to transform hydrolysable salts to non-hydrolysable NaCl is also widely used. As the refiners start to process more and more frequently crudes from the spot market –sometimes meaning heavier crudes or for the same API less quality- many other factors play an important role in the corrosion: H2S, NH3, organic chlorides and light organic acids (LOA).

One of the Process engineer key roles is to study extensively the crude characteristics: i.e. salt content (Ca, Mg, Na), H2S presence and generation at process conditions, presence of LOA, water formation quality, etc. The process conditions: i.e. salt formation temperature, potential for shock condensation, water pH at overhead drum, metallurgy, etc. So, the process engineer is to work closely with chemical suppliers to define the right strategies to tackle down the corrosion.

Strategies may include caustic addition, combination of neutralizers, filmic amines, wash-water and/or combinations of all. The use and efficiency of desalters is also very important: proper adjustment of desalter is a key element to reduce as much as possible the Ca, Mg salts. Depending on refinery configuration, it might be necessary to add some caustic for further corrosion reduction. The addition of caustic will depend on Na balance to Cokers and/or Visbreakers to avoid fast coking, normally happening at levels above 20 ppm Na in the feed to those units.

There are no simple solutions anymore like 30 yrs ago, when using a combination of NaOH and NH3 was good enough for all crudes. General guidelines and examples are given for a variety of light, medium and heavy crudes for different operating conditions and different contaminants.

Paper No. 2b

THE JET FUEL PROBLEM STRIKES AGAIN

Gunter Schnaibel[1]; Andre C. Bellote[2]; Antonio V. S. Castro[3]; Carlos N. Fonseca[3]; Maria E. P. L. Marsiglia[4]; Glaucia A. S. Torres[4]; Karolline Ropelato[5]; Antonio S. Moraes[5]; Fernando Feitosa[5]

[1] REPLAN – Paulinia Refinery, Rod. SP 332, km 132 – Paulinia - Sao Paulo, Brazil, 13140-000
[2] RPBC – Presidente Bernardes Refinery, Praca Marechal Stenio Caio de Albuquerque Lima, 1- Cubatao – Sao Paulo, 11555-900
[3] CENPES – PETROBRAS Research and Development Center, Cid. Universitaria Q.7 – Ilha do Fundao – Rio de Janeiro/RJ, Brazil, 21949-900
[4] PETROBRAS Headquarter, Av. Republica do Chile, 65 – Centro, Rio de Janeiro/RJ, Brazil, 20031-912
[5] ESSS – Engineering Simulation and Scientific Software, Rua Lauro Muller, 116, Botafogo, Rio de Janeiro/RJ, Brazil, 22290-160

Prepared for Presentation at the 2009 AIChE Spring National Meeting, April 26 - 30, Tampa Convention Center, Tampa, FL, Distillation Topical (T8)

Improved Design and Operation in Refinery Distillation - T8001

April, 2009

Unpublished

Abstract

In 1999, REPLAN - Petrobras' biggest refinery - carried out a major turnaround of its Crude Distillation Unit U-200. To keep the costs down, the strategy was to increase unit capacity without changes to the furnaces. In the crude tower most trays were replaced with packing in order to reduce operational pressure and pressure drop allowing a higher vaporization. A new bottom side draw was added, AGO - atmospheric gasoil - directed to FCC charge. This extra side draw allowed the balance of feed and energy absorption between atmospheric and vacuum furnaces by changing the AGO rate.

After start-up, the jet fuel yield decreased to half of its original value with trays. The fractionation between jet fuel and light diesel had deteriorated noticeably, making it impossible to keep the original quality and yield.

A troubleshooting study concluded that the problem was related to liquid distribution on the kerosene and light diesel beds. The main causes were not the distributors themselves, but the collector pans, whose odd design caused flow problems and chimneys' overflow. New collectors were designed and delivered by the internals supplier.

In 2005, unit U-200 went through its first turnaround after revamp and new collectors were introduced. After startup, the column performed as badly as before, sometimes even worse. At this point, two theories came up: 1- some kind of damage or assembly mistake; 2- a poor tower configuration, considering the 1999 revamp had altered the position of the bottom pumparound. In 2006, during a pit stop, the column was opened for inspection and repair but no damages nor assembly errors were found. As part of the plan, a distribution test was conducted to determine the actual liquid distribution and a surprising result was achieved: the liquid distribution was completely skewed. The other beds, that performed normally, presented an even distribution, as expected.

A CFD study focusing chimney arrangement of collector pans was developed to understand the liquid flow dynamics. The results for collector pan sections were input as liquid inlet conditions for the distributors. The effects of liquid non uniformity feeding the distributor were analyzed and best practices in collector/distributor projects were proposed as engineering enhancements for new designs. A three-dimensional (3D) transient, non homogeneous and free surface model (Eulerian-Eulerian approach) designed for flow and geometry improvement of liquid distribution was applied to the phenomenon description.

This paper will present the original and new collector concepts, the field test results, the CFD study and conclusions plus a proposal for a practical fix to minimize the problem. Lessons learned will focus on the correct outlines for any revamp together with correct installation of packing and its associated internals.

KEYWORDS: column internals, trays, packing, jet fuel, kerosene, revamps, CFD.

1- Introduction

REPLAN is Petrobras' biggest refinery. It is located in Southeastern Brazil, Paulinia, Sao Paulo state, and its nominal feed rate is 57,200 m3/d (360,000 bpd). REPLAN has two distillation units, U-200 and U-200A.

The units performed similarly up to 1997 with 24,000 m3/d feed rate each. They were both revamped, U-200A in 1997 to 27,000 m3/d and U-200 in 1999 to 27,200 m3/d, considering a heavier crude slate and higher energy recovery. For the revamps the main constraint was cost, which should be kept as low as possible. For this reason furnaces had to be kept unchanged.

The new designs considered a higher heat recovery in preheat trains and the atmospheric columns had most of their trays replaced by packing in order to reduce flash zone pressure achieving higher vaporization at the same furnace heat release.

After revamp, these units performed well regarding diesel production and quality but both presented jet fuel yield problems, even considering that U-200 design was updated based on U-200A problem study (1997/1998).

The U-200A distillation unit undertook a new revamp in 2003 to 30.000 m3/d where the atmospheric column internals were again replaced, but sticking to packing. After startup its performance complied to design parameters and even exceeded the trays performance. For this revamp a new vacuum furnace was installed. This case was presented in a previous paper by Schnaibel & Marsiglia (2004)[1].

The U-200 unit was kept unchanged until now and Petrobras carried out a troubleshoot study with support of internals supplier trying to fix its problem. This case will be our subject here.

2- Original Problem Analysis

The jet fuel yield reduction was studied through field tests, simulation, gamma & tomography surveys and internal design analysis. Our team concluded that the problem was caused by improper collector design. The following problems were observed:

a) The chimneys were too long and their arrangement favored flow to the side draw, in detriment to the internal reflux rate, which was much higher (withdrawal : internal reflux = 1 : 5);
b) Stiffener plates, which provide mechanical resistance to collectors, created barriers to the liquid flow to the downpipes, raising the liquid level gradient along this path over the collector;
c) The downpipes had not enough area to provide self venting flow, raising the liquid level over them and overall the tray;
d) The downpipes had a high projection over collector's floor and the chimneys were short, leaving only a small room between downpipes top and chimneys edge (90 mm or 3.5 in).

All these problems together caused a high liquid level and gradient on collectors leading to the overflow of the smaller chimneys at least.

The sketch bellow, Figure 1, gives an idea of the arrangement and the photo shows the high liquid level on collector during the water test conducted at 80% of the normal operational flow rate.

Figure 1: Original collector

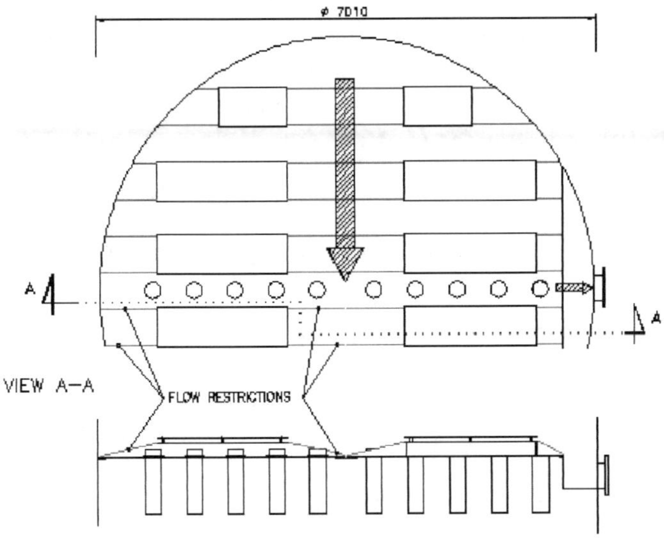

(a) Collector pan arrangement and

(b) The high liquid level on collector.

3- New collector Design

Based on the above reported problems the Kerosene and Light Diesel collectors were redesigned, changing the stiffener plates to a central sump, which provides the required mechanical tray resistance, the downpipes flow area was increased providing low liquid velocity (self-venting) and its projection above collector floor was reduced. Chimneys orientation was kept unchanged to avoid hot work inside the column during turnaround.

Figure 2: New collector tray arrangement

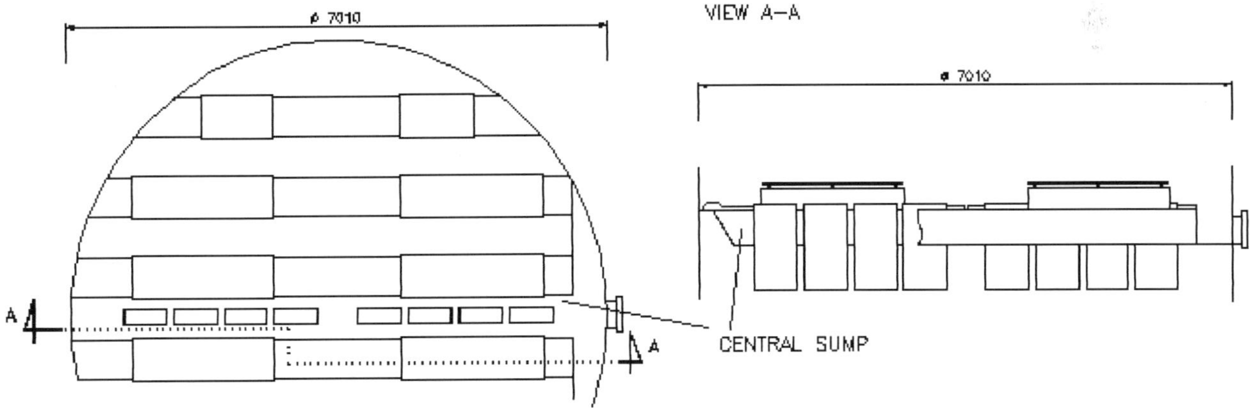

After column startup, the Kerosene yield problem remained. A new troubleshoot study was conducted and our group concluded that the problem could be related to some mechanical damage during start-up or assembly problems or, from a totally different viewpoint, the Light Diesel x Heavy Diesel bed could be running under a very dry condition due to a concept change in the 1999 revamp, where the bottom pumparound was moved down from light diesel draw to heavy diesel draw. Beds running under too low liquid loads are prone to efficiency problems.

In 2006, shortly before the unit pit stop, we performed a radial skin temperature survey at the two suspect beds of the crude tower, namely the kerosene bed and the light diesel bed, assessing the gas region 500 mm above the top of the beds and the liquid region 500 mm below the top of the beds. We measured deltas up to 40°C (70°F) across each elevation; in other words, skewed temperature profiles indicating liquid and gas maldistribution. Graph 1 shows the measured liquid temperature profile for the kerosene bed, before (curves dated 30th January) and after the pit stop (curves dated 15th to 20th February).

Graph 1 – Measured skin temperature profile for the kerosene bed (liquid phase).

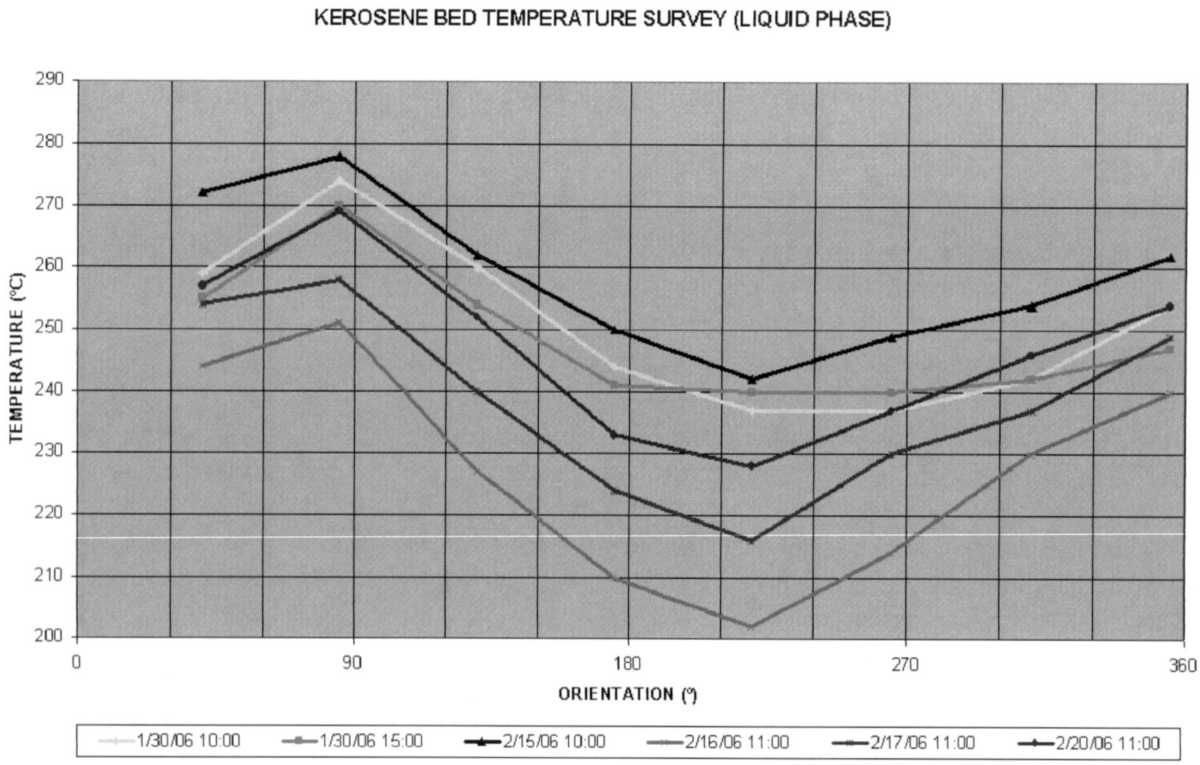

During the unit pit stop, the column was opened to fix the presumable mechanical problems (damage or assembling), but nothing wrong was found. Following our intervention plan, a distribution test, called "wet test" or "water test", was conducted to evaluate the actual liquid distribution. Surprising results came out.

4- Unit Test

Collectors and distributors were checked for levelness and leakages and everything was like design parameters. The next step was the distribution test: the column bottom was filled up with water; the suctions of bottom pumps were connected with the suctions of top pumparound pumps (four six inch hoses connected to filter covers). The top pumparound pumps were put into operation and the flow was controlled to get the regular flow rates over each of the inspected beds. All product drawoffs were blocked, so only internal reflux was flowing during the test (total reflux).

The liquid levels on the distributor troughs were measured; they should be more or less the same over the entire cross section. Measurements were accurate to ± 5 mm, Figure 3.

Figure 3: Distributor Level Measurements

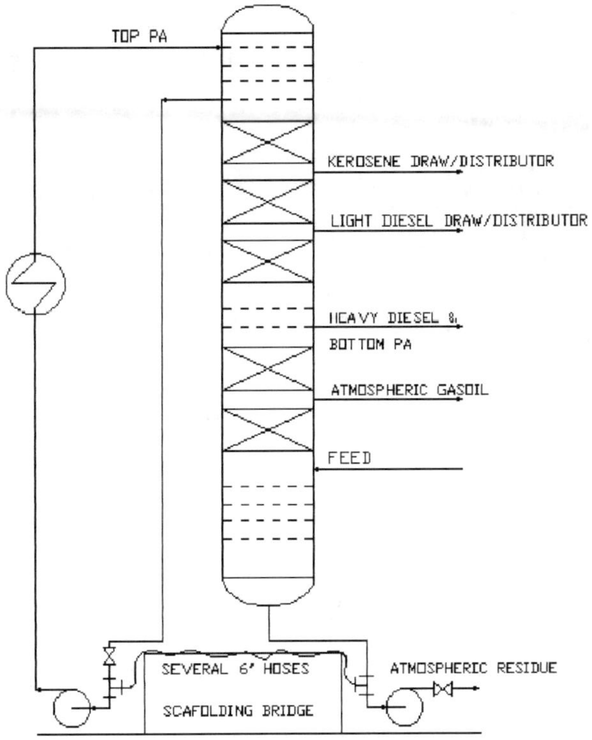

During the water test, the water level on collector pans was checked, assuring that our previous problem, chimney overflow, was eliminated. Nevertheless, an unexpected problem was found.

Beds performing normally presented distributors with liquid heights varying less than 10% around the average (as expected). On the other hand, the two beds with malfractionation presented skewed liquid profiles across the column, as seen in the results below.

Figure 4 shows two beds with good efficiencies, achieving the expected number of theoretical stages. The liquid heights are reasonably uniform. The two different colors of the bars refer to two different flowrates tested.

Figure 4: Measured liquid heights on distributor (Beds with good efficiencies)

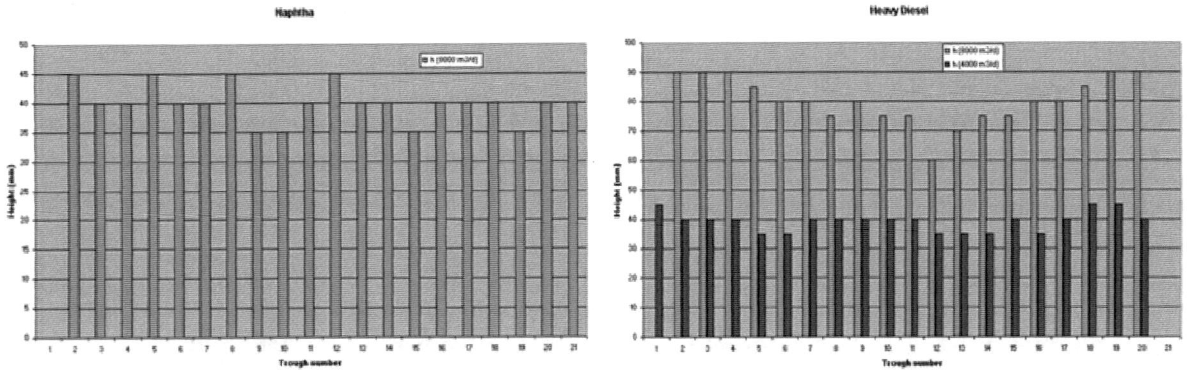

(a) Bed 1 - naphtha bed

(b) Bed 4 - heavy diesel bed.

Figure 5 shows the two beds with low efficiencies, achieving just half of the expected number of theoretical stages. The liquid heights are clearly uneven in each of these beds.

Figure 5: Measured liquid heights on distributor (beds with low efficiencies)

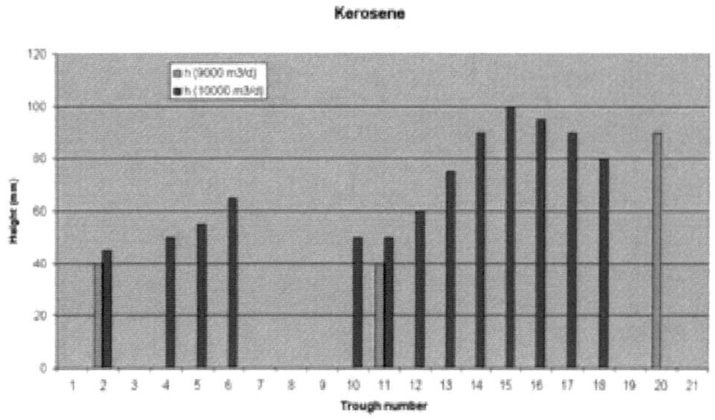

(a) Bed 2 - kerosene bed

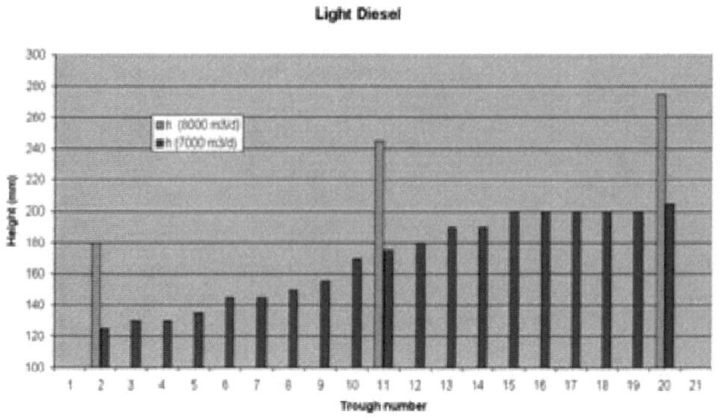

(b) Bed 3 - light diesel bed

Where more data points were collected, a statistical analysis was possible, and the coefficient of variance (Cv) was calculated according to equation 1. The criterion for good distribution is Cv ≤ 10%. If Cv > 10%, values are statistically different. The Cv's reported below statistically express that measured liquid heights are essentially uniform across the naphtha distributor and across the heavy diesel distributor, whereas they were essentially non-uniform across the light diesel distributor and especially across the kerosene distributor.

Equation 1: $Cv = \frac{std\ deviation}{mean} 100\%$

Table 1: Coefficient of Variance for the distributors

	NAPHTHA	HEAVY DIESEL	KEROSENE	LIGHT DIESEL
Cv	8.3%	10.1% / 8.5%	28.2%	17.0%

5- CFD Studies

Computational fluid dynamics (CFD) is a powerful tool applied in design, troubleshooting and development of deep industrial knowledge, leading to better-designed technology and reduced prototyping time. Considering this, CFD techniques were applied to the problem trying to understand the collectors pan liquid phase flow behavior.

The CFD study considered only the new collector design with its central sump and no stiffeners.

This design arrangement revealed an odd behavior because of its long chimneys and their position too close to downpipes or downboxes creating "shadow" regions. The downpipes/downboxes in these shadows had flows far below the average.

Two new arrangements were studied; one shifted (chimneys were shifted 200 mm away from the downboxes) and another rotated, where chimneys were rotated by 90º. Figure 6 presents the geometric alternatives, blue color representing the water inlet, red color indicating the downboxes and green color the draw nozzle. Downboxes and draw nozzle are the liquid outlets. The liquid (water) feeds the system and flows around chimneys to the downboxes and to the central sump.

The mass flow rate at the draw nozzle, presented in Table 3, represents different liquid side draw cases. The total downbox mass flow is the internal reflux, resulting from mass flow balance. The individual downbox flows A, B, C, ... , H are not equal, but affected by the chimney arrangement.

Details of the study were shown in a previous paper by Ropelato et al. (2007)[2]. The standard k-ε turbulence model was used. Comparing the three chimney arrangements applied to our kerosene collector, regarding downbox liquid flow uniformity, the calculated Cv's indicate the best arrangement to be 90° rotated, while the worst, the original one (unfortunately!), as shown under figure 6.

Figure 6: Collector pans.

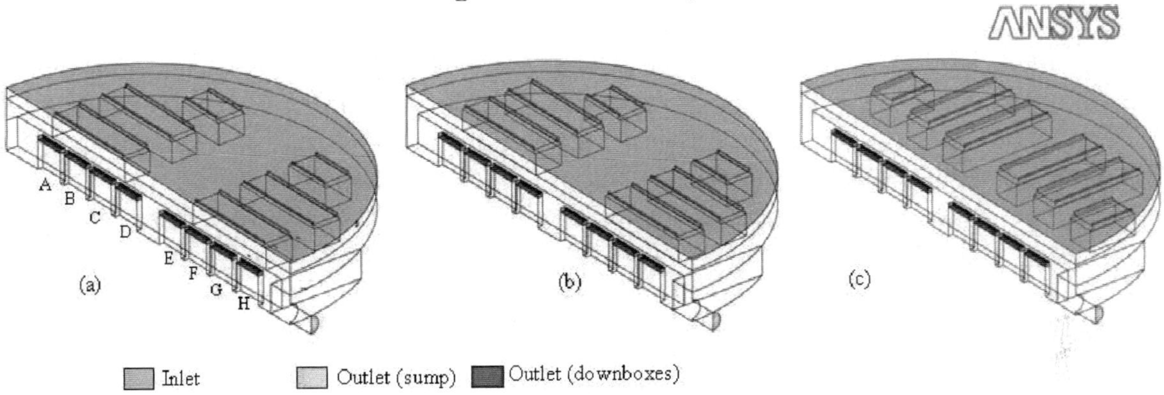

(a) original (b) shifted and (c) 90° rotated

Table 2: CFD results for different chimney arrangements

	ORIGINAL	SHIFTED	ROTATED
Cv	0.1133	0.1069	0.0443

The study was extended to analyze different withdrawal : internal reflux ratios considering all three arrangements. The studied ratios were: 1:5, 1:1 and 5:1, covering the full range of side draws rates in this column.

Table 3: Evaluated side draw conditions.

	Inlet (kg/s)	Outlet (kg/s)	
		Internal reflux	Side draw
Low	95	79.16	15.84
Medium	95	47.5	47.5
High	95	15.84	79.16

Table 4 presents an overview of simulated results, with crosses indicating the best options. For high drawoffs, a total drawoff pan is recommended, considering all simulated arrangements failed to achieve acceptable distribution. These results may be conservative, but can be generalized and considered rules of thumb for choosing the layout of the chimneys as well as the number of withdrawals.

Table 4: Results overview

	1 Draw nozzle			2 Draw nozzles		
	Original	Shifted	Rotated	Original	Shifted	Rotated
Low			X			
Medium					X	X
High						

After this the distributor was also simulated. The simulation accounted for the flow field generated in the collector pan section. Effects of a non-uniform feed to the distributor were investigated.

Details of distributor geometry are presented in Figure 7 (a). The red color represents the liquid that comes from the previous section (the collector pan section) and the green color indicates the details of the liquid outlets. Considering the huge geometric scale difference, the grid around the orifices has to be very refined to have a good numerical approximation, but the number of nodes (points of calculation) would be higher than 2 million, leading to a huge calculation time.

This numerical limitation rendered it impractical; therefore, a homogeneous single-phase approach was developed and used. This approach made it impossible to accurately predict the liquid level. However, a qualitative analysis could be done to analyze the influence of a non uniform liquid feed upon distributor operation.

Figure 7: Liquid distributor

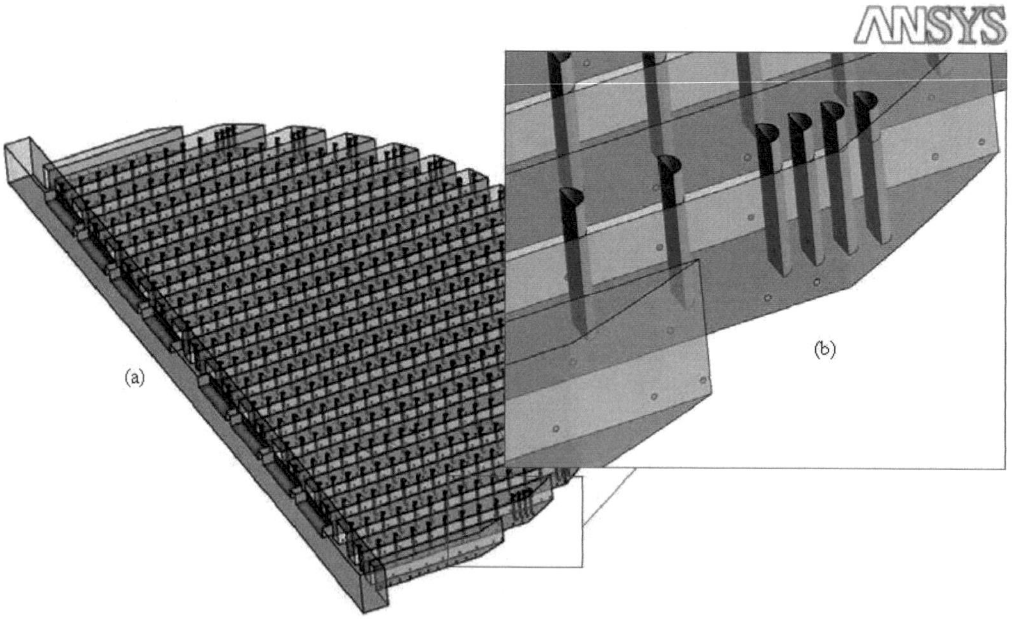

(a) Liquid distributor (b) Detail of liquid outlet region.

Two different inlet conditions were studied. The first one considered the non-uniform fluid distribution generated by the original collector pan geometry, where each downbox (inlet) had a different mass flow. The second case applied the mean flow to each downbox, equally split.

Figure 8 shows the achieved results. Figure 8 (a) shows that the bad flow distribution did not allow the liquid velocity profile to become uniformly distributed, even after a complete residence time. For the condition of even mass flow to all downboxes, a uniform profile was reached, as seen in Figure 8 (b).

Figure 8: Distributor velocity profiles

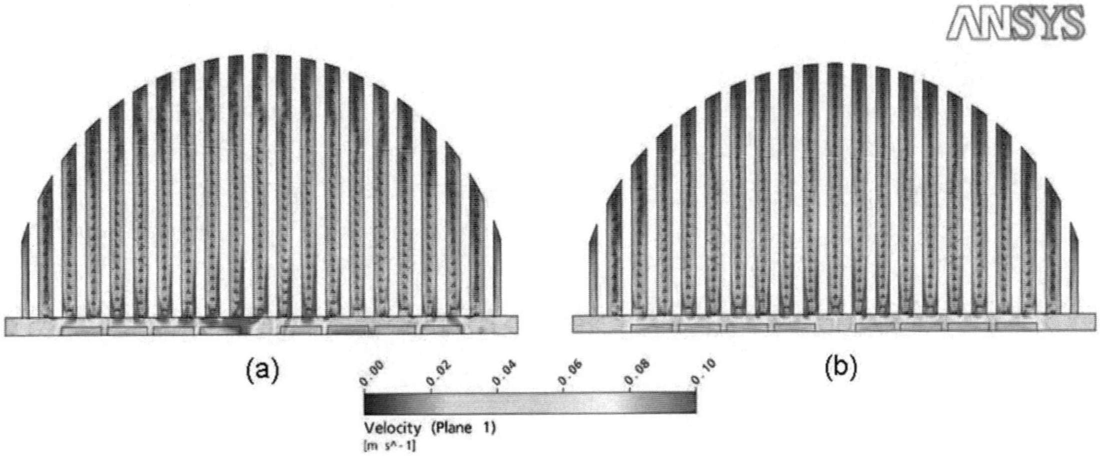

(a) Different liquid mass flows in each downbox (b) Even liquid mass flows in each downbox

Although several approaches were tried, all previously described, plus tilted collectors, the simulation results couldn't reproduce the measured profiles. Our group discussed the possible causes and concluded that probably some of the initial assumptions for the simulation were untrue.

Example of questionable assumptions:

a) Even liquid distribution from the bed above reaching the collector. Certainly the liquid leaving the kerosene bed to the light diesel collector couldn't be even;
b) Distributor levelness. Although confirmed by the water test, it's possible to have a small deviation, which could be a problem in 7 meter (23 feet) diameter column.
c) Even gas distribution coming from the bed below. If liquid distribution is skewed, so would the gas distribution be, as gas would preferentially flow in the drier sections of the beds. However, this cannot be the cause, but rather a consequence of our maldistribution problem; hence it doesn't explain why the CFD simulation failed to reproduce measured profiles.

6- Collector improvements

The CFD study was also used to analyze some practical fixes that could minimize the collector gradient and bring each downbox flow closer to the average. The alternative which presented the best results was the inclusion of new weirs on each side of the central sump. This obstruction would raise liquid level on collectors (nowadays there is enough room) and reduce the liquid gradient along the sump, allowing a more uniform flow to each downpipe or downbox, as can be seen in figure 9.

Besides, for both collectors the existing chimneys should be rotated by 90º or at least shifted away from central sump.

Examining the history of these collectors, it can be said that we migrated from a high pressure drop situation in the original collector pan, to quite the opposite situation in the new collector, where liquid distribution became helplessly attached to the poor initial geometry of that pan. No wonder the proposed practical fix is the inclusion of a weir - an obstruction intended to even out pressure drop across the pan.

The original collector pans had factors (a)+(b)+(c)+(d), listed in section 2 above, contributing for high liquid gradient, which culminated in chimney overflow (recalling: a - chimneys too long and transverse to main flow, b - stiffener plates, c - non-self-venting downpipes, d - taller downpipes). With the new collector pan in 2005, all factors but (a) were eliminated. Thus, we got rid of the original collector's overflow problem, but instead… we stepped right into a maldistribution problem, still caused by the collector – more precisely, by factor (a), which was not eliminated!

Figure 9: Proposed practical fix

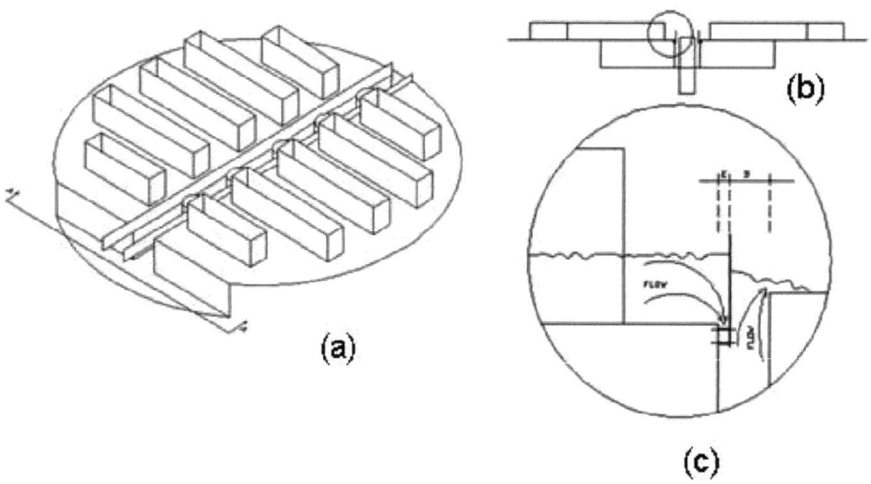

(a) Collector pan view, (b) Position of the new weirs and (c) Detail of the new weirs.

7- Distributor concerns

The distributors used in the column are very simple; the downpipes/downboxes feed the liquid directly into a central main parting box without any sealing, and the central main box feeds troughs directly through windows on its sides.

In three beds of the column this distributor design is working reasonably well (all three are fed by collectors with chimneys rotated by 90° and a central sump); but there are the reported problems in other two collectors (Kerosene and Light Diesel).

Evaluating the design of higher quality distributors on the market, it could be seen that most of them have the downpipes sealed, use a pre-distributor inside the main distributor box and have some kind of device to reduce splashing. Another point is that most of them have the troughs assembled below the main box and they are fed through calibrated orifices so that each trough receives the right amount of liquid, minimizing distribution problems.

Checking literature design recommendations it could also be seen that for columns larger than 3,5 meters in diameter the distributor should consider the use of two main parting boxes, our column is 7 meters in diameter and has only one main box.

8- Acknowledgements

Cyro Costa Junior, Replan Engineering manager (who retired this year).
Gilberto Aoqui, Luis Augusto Escuciato and members from Replan Distillation Units
Magno Schiavolin and members from Replan Optimization and Logistic Sectors
Edneu Jatkoski and members from Replan Equipment Inspection team
Marisa Makiko Matsuda, Replan Optimization manager
CENPES support through employees, equipment (cameras) and CFD software, contracts and personnel.

9- Conclusions

Some great lessons were learned from the jet fuel problem.

Number 1: **Mechanical and process design must be in line to get good results.**

We saw this in the 1999 U-200 case: Overlooked details such as those chimney reinforcements or chimney orientation created unexpected barriers to liquid flow. Final result: Kerosene yield reduced to half.

There is a second reading from Lesson 1: packing is vulnerable to a lot more factors than trays. Trays naturally redistribute liquid and vapor at each one of them, so random errors do not add up to a systematic error. In this sense, trays are more robust than packing. Therefore, we restate Lesson number 1 as: **PACKING solutions may fail if mechanical and process designs are not in line.**

Number 2: **Liquid distribution was the key issue in the case presented in this paper.**

Liquid distribution is one of the main causes of packing solution failures and in most of the related cases the villain was the distributor. In our case the specific cause of liquid maldistribution lied not on the distributors themselves, but on collectors.

Number 3: **Use of CFD studies**

CFD modeling is a great tool to get better understanding of hydraulic flow problems and it is able to bring up problems that are not straightforward for people who don't deal with troubleshooting daily. But a CFD simulation, as any other simulation, depends strongly on the data, modeling quality and initial assumptions. In our case the simulations pointed the main causes but we weren't able to reproduce test measurements, probably because of inadequate initial assumptions.

Number 4: **Collector and Distributor design**

The CFD studies confirmed that liquid distribution must start as even as possible. If the collector delivers an uneven feed to the distributor, it is incapable to correct the problem even if the residential time is high. In resume **"If it starts the wrong way, it will keep that way…"**. In this line we consider very important the test of distributors at vendor shop, remember that the test must mimic the real feeding system in order to reproduce the field situation as well as possible.

Some best practices in collector pan projects were proposed as an engineering feature on new designs. The results of this study pointed the need of an engineering analysis for each column internals selection, taking into account general operation condition, especially for packed columns. The optimum performance and desired operating ranges can only be achieved when tower packing/internals are precisely designed and carefully selected for individual cases.

REFERENCES

1) Schnaibel, G., Marsiglia, M. E. (2004). **The Recurring Jet Fuel Problem (and what it took to solve it).** AIChE 2004 Spring National Meeting, New Orleans, LA, USA
2) Ropelato, K., Rodrigues, R., Bellote, A. C., Schnaibel, G., Torres, G. A. S., Marsiglia, M. E. P. L., Fonseca, C. N., Castro, A. V. S. (2007). CFD Study for chimney tray design in crude fractionators. AIChE 2007 Annual Meeting, Salt Lake City, Utah, USA
3) Kister, H. Z., Distilation – Operation, MacGraw-Hill, 1992
4) Wilcox, D. C. (2000) Turbulence Modelling for CFD. DCW Industries, pp. 314.
5) M. Heggemann, S. Hirschberg, L. Spiegel and C. Bachmann (2007). CFD Simulation And Experimental Validation Of Fluid Flow In Liquid Distributors. Chemical Engineering Research and Design. Vol 85 (A1) pages 59–64.
6) H. Z. Kister, B. Blum, T. Rosenzweig (2001). Troubleshoot chimney trays effectively. Hydrocarbon Processing. April 2001.

Attachment 1 – Original columns sketch (U-200)

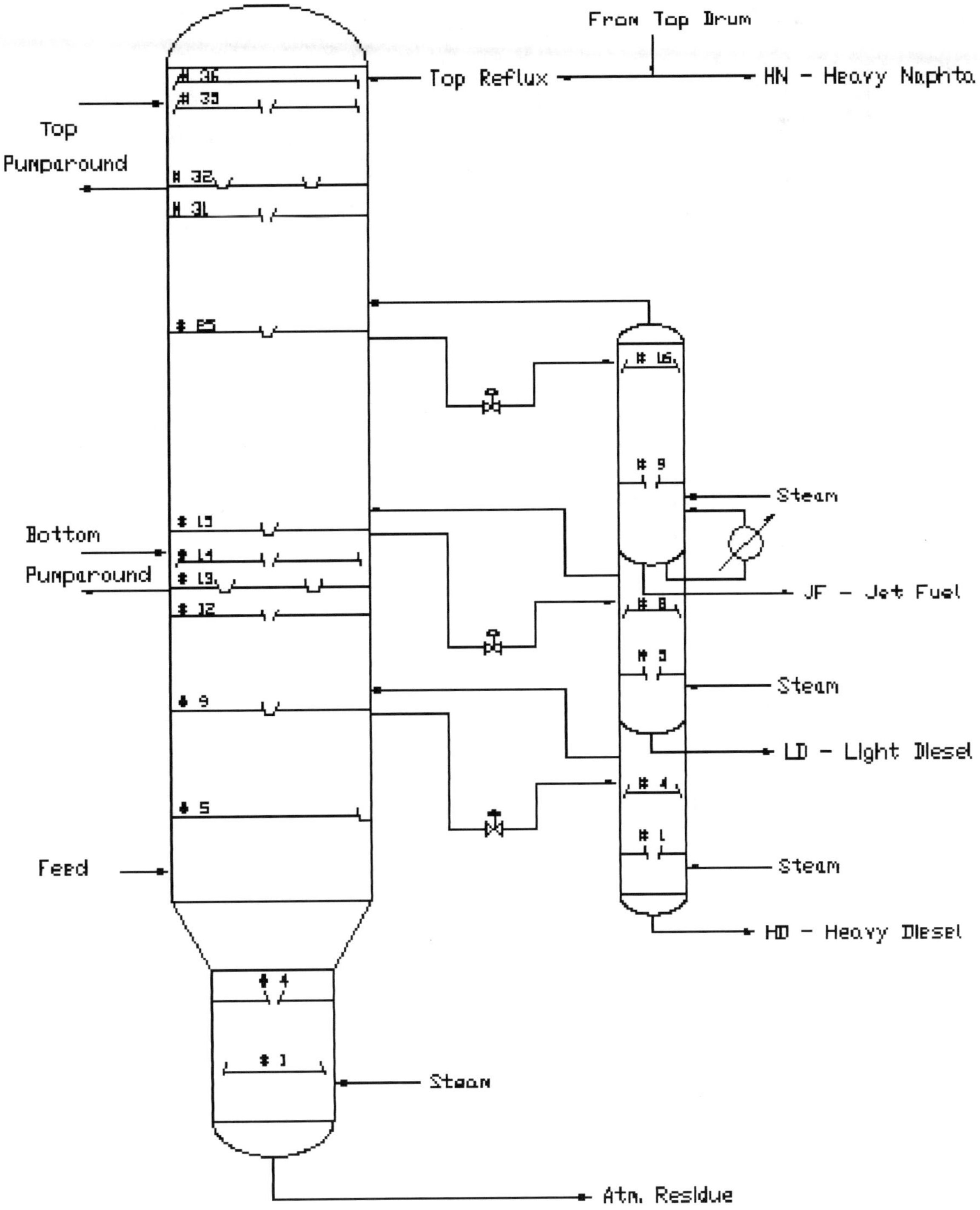

Attachment 2 – U-200 Crude Column after 1999 revamp

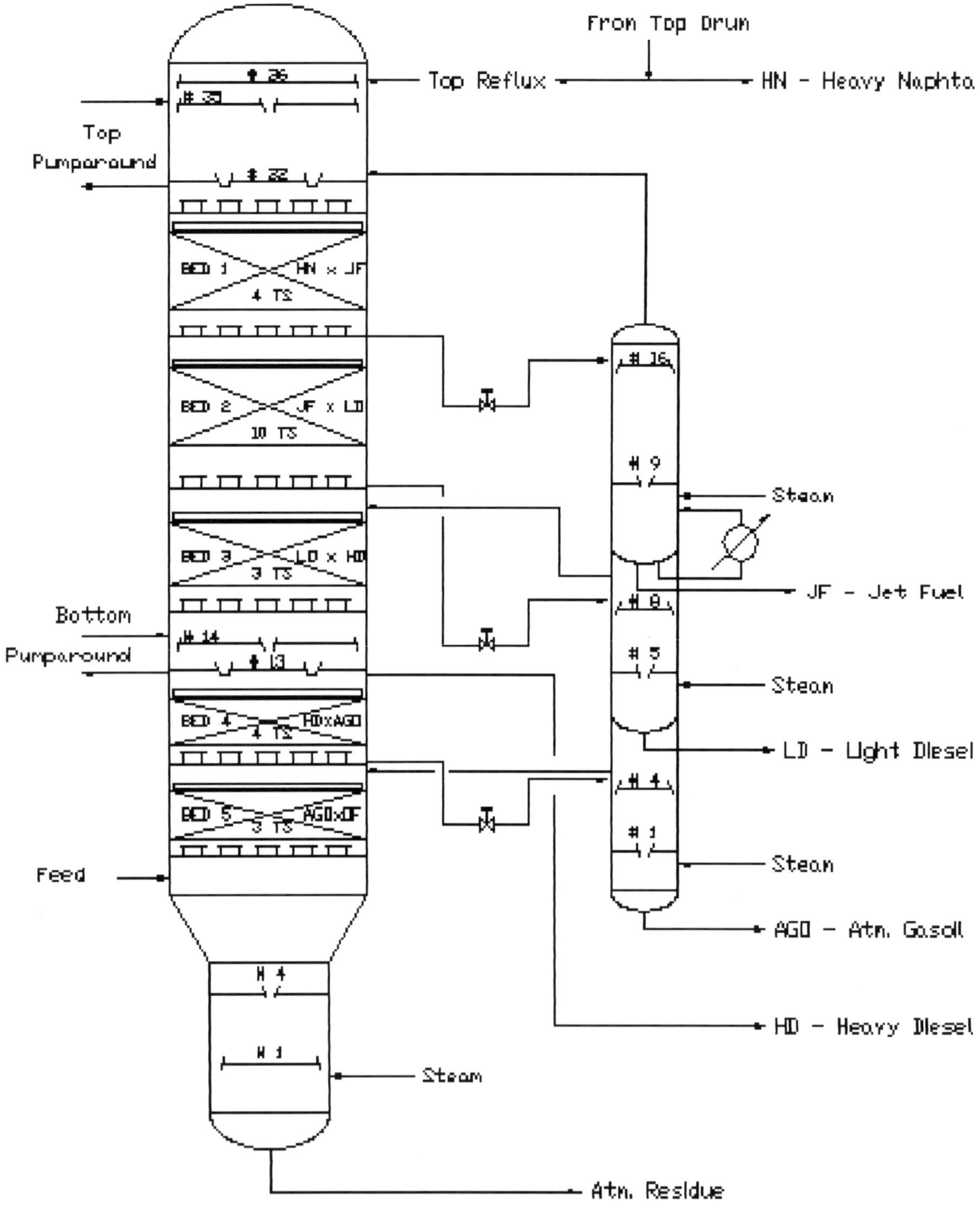

26c What Went Wrong in the Design of the Total Draw off Tray and the Thermosiphon Loop on a Crude Oil Stabilizer Column?

Gabriel T. Fernandez, Process & Control Systems Department, Saudi Aramco, 12921, Dhahran 31311, Saudi Arabia

This paper studies the design of the total draw off tray by the column Tray Vendor and the design of the thermosiphon loop including reboilers, by an Engineering Design Contractor. Though the Tray Vendor and the Engineering Design Contractor were both right in their designs, the system design was in error and would have resulted in unstable thermosiphon loop operation. This paper and presentation will discuss the individual designs and the recommendation made by the client to correct this serious error during the design review. The change was accepted by all parties and was constructed as recommended.

26d Demystifying the Zero Entrainment Myth

Mario Roza[1], Markus Duss[1], **Mark W. Pilling**[2], and SM Wong[2]. (1) Applications & Process Technology, Sulzer Chemtech AG, Switzerland, Switzerland, (2) Sulzer Chemtech USA, Inc., 4019 S. Jackson Ave., Tulsa, OK 74107

In the last few years the importance and feasibility of vacuum tower inlet devices having very little or no entrainment to the column slop draw has been very strongly promoted. Entrainment from the flash zone is only one aspect which determines the quality and yield of vacuum gas oil which can be obtained for given crudes.

In the presentation all aspects influencing the vacuum gas oil quality are considered. Starting from the flash we will look at entrainment from the inlet device, de-entrainment characteristics of the internals in the wash section and entrainment from the wash oil distributor.

For each of these aspects experimental data, CFD analysis and data obtained from actual columns will be shown. Results from modeling of different radial and tangential inlet devices as well as a modeling of entrainment from spray nozzle distributors will be presented.

Finally the importance of all of these factors determining the maximum yield for a given quality of vacuum gas oil in particular the effect of entrainment of the inlet device are put in perspective.

26e Crude Vacuum Tower Wash Section - a New Design Approach

Carlos Trompiz[1], Garry Jacobs[2], and Henry Z. Kister[2]. (1) Fluor, Houston, TX 92698, (2) Fluor Corporation, 3 Polaris Way, Aliso Viejo, CA 92698

Details on vacuum tower wash section - a new way to design it.

EMPTY SPRAY SECTION RESULTS AFTER START-UP: HEAT TRANSFER AND ALSO FRACTIONATION

Ana Lídia Wild[1]
Elizabeth Ferreira da Fonseca[2]
Glaucia Alves da Silva Torres[3]
Marcos Silva de Souza[1]
Silvia Waintraub[2]

[1]PETROBRAS/REFAP S.A.
[2]PETROBRAS/RESEARCH CENTER
[3]PETROBRAS/HEADQUARTERS

Rio de Janeiro
Brazil

Prepared for Presentation at the AIChE 2009 Spring National Meeting
Tampa, April 26 - 30

March/2009

Unpublished

Abstract

The vacuum tower of REFAP refinery, located in southern Brazil, was revamped with a too deep modification: packing were removed from the light gas oil pumparound and slop wax wash sections. At the same time the light gas oil x heavy gas oil fractionation section remained with the spray distributor and no internals, as an additional heavy gas oil pumparound section.

This tower started-up in February 2007 and has been working well since then.

A test run was performed in order to evaluate the light gas oil pumparound empty spray section. The influence of hot light gas oil reflux ratio in the separation quality of light gas oil x heavy gas oil was also observed at each step of the test.

The results of these plant tests are going to be presented in this paper.

Introduction

Petrobras has been studying vacuum distillation tower configurations in the past few years through CFD analysis and plant tests (Barbabela *et alii*, 2004, Waintraub *et alii*, 2003, Waintraub *et alii*, 2005).

Recent papers published by Petrobras related to empty spray sections in vacuum towers showed their viability and effectiveness in heat transfer applications.

In the 2003 paper, some characteristics were showed:

- The heat transfer for sub cooled liquid reflux occurs at the bottom of the region as at the upper section temperature is constant;
- The heat transfer in empty spray sections depends on both liquid and vapor rates;
- When increasing the vapor rate, the height necessary to a given exchange is higher;
- There is a maximum value of the vapor rate (C-factor) above which the stability of the spray cone shape is lost.

In the 2005 paper, the revamp of a vacuum tower with three empty spray sections was presented, and in this work, the results after the start up proving the viability and effectiveness of the empty spray sections in this tower are shown.

The main objective is to reduce the column pressure drop with the use of empty spray sections in order to increase profitability through a deeper cut with less energy consumption in the vacuum heater.

Previous to 2007, test runs in two empty spray sections in Brazilian refineries were performed: one in REFAP refinery due to a packing bed that was removed remaining only the spray distributor and another in REDUC refinery that was designed by Badger Limited in 1975. The knowledge obtained through these studies allowed Petrobras to revamp REFAP vacuum tower in that year.

Originally REFAP topping unit was designed in 1968 to process 7,500m³/d of crude. Its vacuum distillation started up in 1970 with a feed rate of 3,700m³/d reduced crude.

In 1991 both topping and vacuum units were revamped to 12,000m³/d and 5,000m³/d, respectively. The existing Pall ring packing at the LVGO pumparound, HVGO pumparound, wash bed and slop wax pumparound were replaced for IMTP packing and a LVGO x HVGO fractionation bed was installed in order to allow the LVGO to be incorporated to the diesel pool.

With the start-up of REFAP second topping unit, both LVGO and HVGO were sent to FCC unit and the LVGO x HVGO fractionation bed was removed in 1998. The spray distributor remained inside the tower, with no packing below it, as an additional heavy gas oil pumparound section. This was the vacuum tower configuration for ten years, as presented in Figure 1a.

Due to increasing Brazilian heavy naphtenic crude oils processing, in the 2007 turnaround project, the vacuum tower had to be revamped to 5,600m³/d. In order to maintain the same cut point without modifying the vacuum heater, empty spray sections, substitution of random packing for higher capacity ones and new inlet feed device were designed to reduce flash zone pressure.

This tower new configuration started-up in February 2007 and has been working well since then, having even processed feed rates as high as 6,000m³/d. A sketch is presented in Figure 1b.

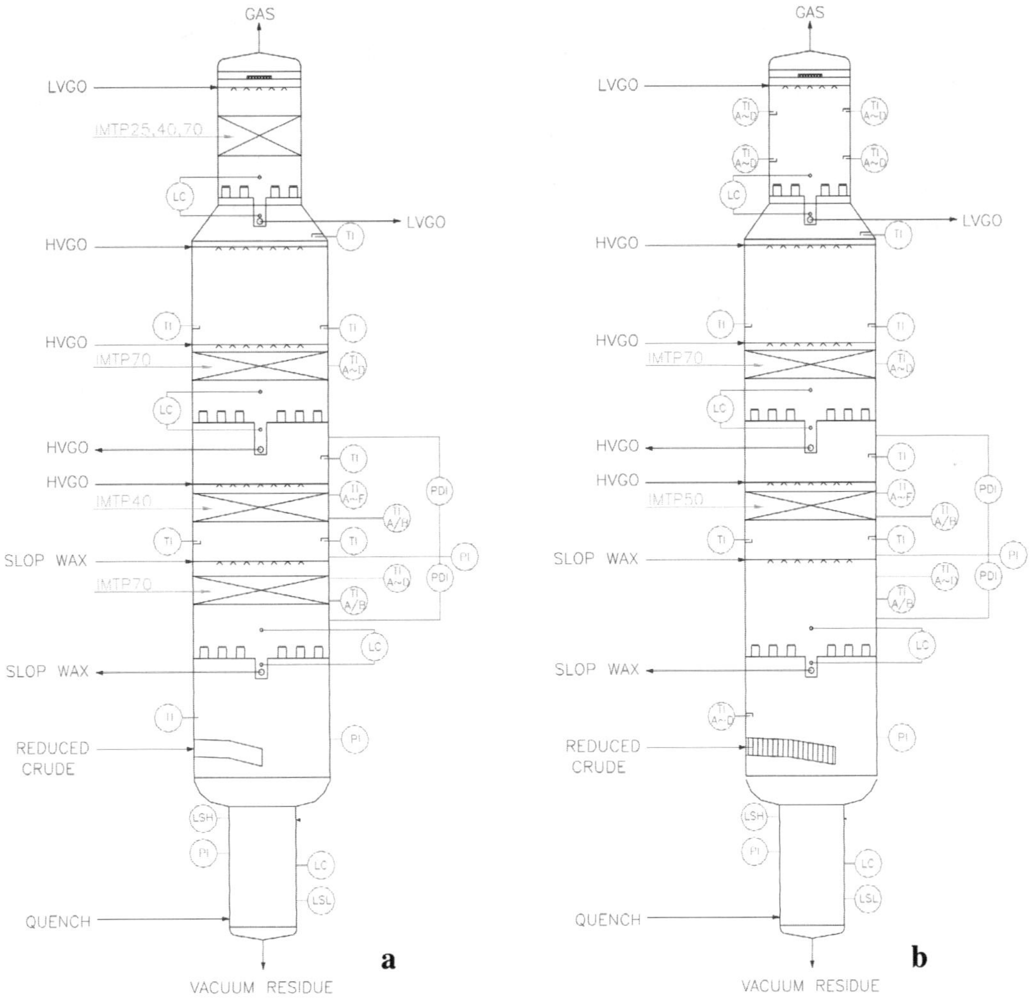

Figure 1 **REFAP Vacuum Tower (a) before and (b) after 2007 Revamp**

The 2007 REVAMP

The vacuum tower was revamped with a too deep modification consisting of:

1. Packing removal from the LVGO pumparound and substitution of the spray distributor. The number of nozzles was increased from nineteen to thirty in order to guarantee good distribution, Figure 2a. Sixteen thermocouples (eight for the liquid phase and eight for the vapor phase) distributed in two elevations were installed, Figure 2b. The temperatures are sent directly to a computer system to monitor the process on line.

A b

Figure 2 **Revamp pictures: (a) new LVGO distributor; (b) thermocouples**

2. Packing was removed from the slop wax pumparound, remaining only the spray distributor.

3. Substitution of wash oil random packing for one with higher capacity (IMTP 40 for IMTP 50).

4. Installation of a new 270° inlet vapor horn from Sulzer, Figure 3b, in place of the old model 90°, Figure 3a.

Previous CFD studies (Waintraub *et alli*, 2005) concluded that the vapor velocity was too high and unevenly distributed in the region below the slop wax collector causing liquid entrainment to the section above. Extremely high local values were found indicating that the use of empty spray section was not practicable, even with a not very high global C-factor (0.30 ft/s). The design of an empty spray section at the slop wax pumparound indicated that the installation of a new feed distributor was required.

5. Light gas oil x heavy gas oil fractionation section remained with the spray distributor and without internals, as an additional heavy gas oil pumparound section.

Design C-factors were obtained using Petrobras process simulator PETROX. For the LVGO section, the C-factor was below Petrobras established limit to apply empty spray sections.

a

b

Figure 3 **Vapor horn pictures: (a) old; (b) new from Sulzer**

Test Run Description

A test run was performed in order to evaluate the light gas oil pumparound empty spray section.

The test was divided in seven steps as summarized in Table 1. The first four steps were done with the crude tower optimized. Steps 5 to 7 were done with AGO purposely degraded from the atmospheric to the vacuum tower, in order to maximize the vapor load to the LVGO empty section.

To evaluate the separation quality between LVGO and HVGO in the empty spray section where the fractionation bed existed, LVGO was sent to the spray distributor instead of HVGO. The influence of the reflux ratio in the separation was observed increasing the LVGO hot reflux flow rate to the distributor.

In each step, the HVGO pumparound rate was reduced allowing more vapor to the section above. At the same time the LVGO pumparound rate was increased, in order to have more condensation yielding more liquid (LVGO product and LVGO hot reflux).

Table 1: Test run steps

STEP	1	2	3	4	5	6	7
REDUCED CRUDE (m^3/d)	BASE	BASE	BASE	BASE	+ 300	+ 300	+ 300
LVGO PUMPAROUND (m^3/d)	BASE	BASE	+ 300	+ 600	+ 200	+ 200	+ 600
LVGO HOT REFLUX (m^3/d)	0	350	700	1000	0	450	1000
HVGO PUMPAROUND (m^3/d)	BASE	- 300	- 600	- 700	+ 600	BASE	- 700

Radial windows were opened in the tower insulation at the LVGO empty spray section to allow temperature measurements with a contact pyrometer. The windows were located at the same heights as the sixteen internal thermocouples (eight for the liquid phase and eight for the vapor phase). The purpose of these locations was to allow the comparison of skin and internal temperatures.

The measurements of these skin temperatures together with the internal thermocouples values helped to determine the distribution profile (both liquid and vapor) and also the height of the void space necessary to exchange the desired heat amount. In Figure 4 a sketch of the empty space region is presented.

After unit stabilization in each step, product samples were collected and also skin temperatures were measured with a pyrometer.

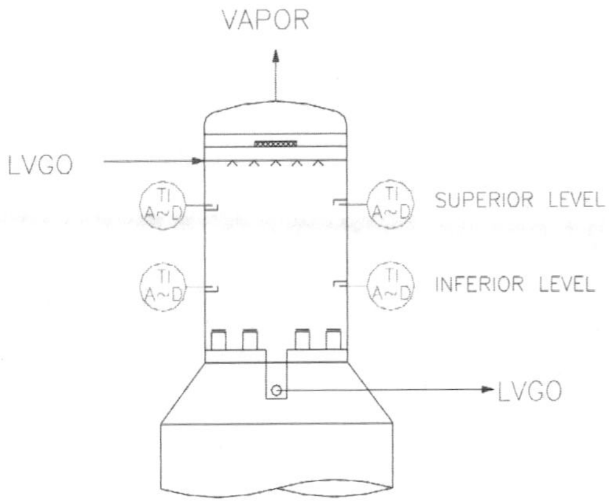

Figure 4 **Empty spray section sketch**

Results and Analysis

Furnace Outlet Temperature and Vacuum Tower Pressure Drop

Due to the removal of two random packing sections, the column total pressure drop had a sharp decrease, allowing the reduction of the outlet furnace temperature, with consequent significant energy savings.

It is interesting to point out that the new condition of 7 mmHg total column pressure drop can even be reduced further if a new design replacing the two existing random beds for structured packing is made.

The average operating pressure and temperature before and after the 2007 revamp are presented in Table 2.

Table 2: Average operating pressure and temperature

PROPERTY	BEFORE REVAMP	AFTER REVAMP
TOP PRESSURE (mmHg)	BASE	-1
FLASH ZONE PRESSURE (mmHg)	BASE	- 10
COLUMN PRESSURE DROP (mmHg)	16	7
FURNACE OUTLET TEMPERATURE (°C)	BASE	-30
FLASH ZONE TEMPERATURE (°C)	BASE	-30

C-Factor

The different steps consisted basically in actions to increase the vapor load to the LVGO empty spray section, that means increase the C-factor.

It was found that even with a sharp increase in C-Factor (over twofold), the heat transfer was accomplished.

Table 3: Empty spray section C-Factor

STEP	1	2	3	4	5	6	7
REDUCED CRUDE (m^3/d)	BASE	BASE	BASE	BASE	+ 300	+ 300	+ 300
LVGO PUMPAROUND (m^3/d)	BASE	BASE	+ 300	+ 600	+ 200	+ 200	+ 600
LVGO HOT REFLUX (m^3/d)	0	350	700	1000	0	450	1000
HVGO PUMPAROUND (m^3/d)	BASE	- 300	- 600	- 700	+ 600	BASE	- 700
INCREASE IN C-FACTOR (%)	BASE	26	76	124	17	59	116

Temperature Profile

Empty Spray Section Superior Level Temperatures

In Table 4, liquid, vapor and skin temperatures at the superior level of the LVGO empty spray section are shown as well as the maximum difference between them. Liquid, vapor and skin average temperatures and their differences are presented in Table 5.

It can be seen that:

- There is a very good liquid distribution in all test run steps;
- There is a good vapor distribution for steps 1, 2 and 5. For higher vapor loads (steps 3, 4, 6 and 7), the vapor distribution was not good;
- It could be said that for steps 1, 2 and 5, almost all the heat transfer occurred from the collector pan to the superior thermocouple level. On the other hand, for steps 3, 4, 6 and 7, it is necessary greater height to accomplish the desired heat transfer, but still below the spray distributor, where there is still more void space. There are no temperature measurements in this region, but we can say that the existing height is sufficient, observing the low vapor overhead temperatures in all steps;
- There is very low difference between the wall and internal temperatures, being the skin values closer to the liquid ones.

Table 4: Empty spray section superior level temperatures

STEP	1	2	3	4	5	6	7
REDUCED CRUDE (m^3/d)	BASE	BASE	BASE	BASE	+ 300	+ 300	+ 300
LVGO PUMPAROUND (m^3/d)	BASE	BASE	+ 300	+ 600	+ 200	+ 200	+ 600
LVGO HOT REFLUX (m^3/d)	0	350	700	1000	0	450	1000
HVGO PUMPAROUND (m^3/d)	BASE	- 300	- 600	- 700	+ 600	BASE	- 700
VAPOR OVERHEAD TEMP (°C)	38	47	48	54	43	56	65
Liquid Temp Thermocouple 1 (°C)	48	50	53	56	51	56	66
Liquid Temp Thermocouple 2 (°C)	48	50	53	68	51	58	77
Liquid Temp Thermocouple 3 (°C)	48	51	54	56	52	56	62
Liquid Temp Thermocouple 4 (°C)	48	51	61	71	52	67	73
Liquid Thermocouple Max ΔT (°C)	0	1	8	15	0	11	15
Vapor Temp Thermocouple 1 (°C)	52	54	60	61	54	69	99
Vapor Temp Thermocouple 2 (°C)	50	58	68	91	54	74	114
Vapor Temp Thermocouple 3 (°C)	56	60	77	76	58	77	86
Vapor Temp Thermocouple 4 (°C)	54	59	111	116	57	119	110
Vapor Thermocouple Max ΔT (°C)	6	6	51	56	4	50	28
Skin Temp Point 1 (°C)	47	48	-	-	50	52	60
Skin Temp Point 2 (°C)	46	46	-	-	49	55	89
Skin Temp Point 3 (°C)	46	46	-	-	50	53	59
Skin Max ΔT (°C)	1	2	-	-	1	3	30

Table 5: Empty spray section superior level average temperatures and differences

STEP	1	2	3	4	5	6	7
REDUCED CRUDE (m^3/d)	BASE	BASE	BASE	BASE	+ 300	+ 300	+ 300
LVGO PUMPAROUND (m^3/d)	BASE	BASE	+ 300	+ 600	+ 200	+ 200	+ 600
LVGO HOT REFLUX (m^3/d)	0	350	700	1000	0	450	1000
HVGO PUMPAROUND (m^3/d)	BASE	- 300	- 600	- 700	+ 600	BASE	- 700
Liquid Average Temperature (°C)	48	51	55	63	51	59	69
Vapor Average Temperature (°C)	53	58	79	86	56	85	102
Skin Average Temperature (°C)	46	47	-	-	50	53	58
Vapor – Liquid Temperature (°C)	5	7	24	23	4	25	33
Skin – Liquid Temperature (°C)	1	4	-	-	2	6	2
Skin – Vapor Temperature (°C)	7	11	-	-	6	31	35

Empty Spray Section Inferior Level Temperatures

In Table 6 liquid, vapor and skin temperatures at the lower level of the LVGO empty spray section are shown as well as the maximum differences between them. Liquid, vapor and skin average temperatures and their differences are presented in Table 7.

Table 6: Empty spray section inferior level temperatures

STEP	1	2	3	4	5	6	7
REDUCED CRUDE (m^3/d)	BASE	BASE	BASE	BASE	+ 300	+ 300	+ 300
LVGO PUMPAROUND (m^3/d)	BASE	BASE	+ 300	+ 600	+ 200	+ 200	+ 600
LVGO HOT REFLUX (m^3/d)	0	350	700	1000	0	450	1000
HVGO PUMPAROUND (m^3/d)	BASE	- 300	- 600	- 700	+ 600	BASE	- 700
Liquid Temp Thermocouple 1 (°C)	48	51	58	64	52	65	83
Liquid Temp Thermocouple 2 (°C)	51	53	62	53	53	66	95
Liquid Temp Thermocouple 3 (°C)	49	53	74	76	53	73	89
Liquid Temp Thermocouple 4 (°C)	51	53	82	84	54	89	87
Liquid Thermocouple Max ΔT (°C)	3	2	25	31	2	24	12
Vapor Temp Thermocouple 1 (°C)	103	108	115	125	105	129	139
Vapor Temp Thermocouple 2 (°C)	92	78	89	116	83	91	140
Vapor Temp Thermocouple 3 (°C)	106	117	135	124	104	131	139
Vapor Temp Thermocouple 4 (°C)	79	81	96	100	94	107	109
Vapor Thermocouple Max ΔT (°C)	26	39	46	25	22	40	32
Skin Temp Point 1 (°C)	49	52	-	-	53	69	73
Skin Temp Point 2 (°C)	51	57	-	-	54	59	86
Skin Temp Point 3 (°C)	49	53	-	-	54	84	87
Skin Max DeltaT (°C)	2	5	-	-	1	25	14

Table 7: Empty spray section inferior level average temperatures and differences

STEP	1	2	3	4	5	6	7
REDUCED CRUDE (m^3/d)	BASE	BASE	BASE	BASE	+ 300	+ 300	+ 300
LVGO PUMPAROUND (m^3/d)	BASE	BASE	+ 300	+ 600	+ 200	+ 200	+ 600
LVGO HOT REFLUX (m^3/d)	0	350	700	1000	0	450	1000
HVGO PUMPAROUND (m^3/d)	BASE	- 300	- 600	- 700	+ 600	BASE	- 700
Liquid Average Temperature (°C)	50	53	69	69	53	73	88
Vapor Average Temperature (°C)	95	96	109	116	96	114	132
Skin Average Temperature (°C)	50	54	-	-	54	71	82
Vapor – Liquid Temperature (°C)	45	43	40	47	44	41	43
Skin – Liquid Temperature (°C)	0	- 1	-	-	- 1	3	6
Skin – Vapor Temperature (°C)	45	42	-	-	43	44	50

We may say that:

- The heat transfer occurs from the collector pan to the section above;
- In the present height, the heat transfer is still being accomplished;
- It is necessary a height above this level to transfer the heat amount in all steps.

Flash Zone Temperatures

The feed inlet device performance was evaluated through thermocouples installed equally spaced inside the tower approximately 1750mm above the upper surface of the vapor horn. In Table 8 it can be seen that there is a very good initial vapor distribution in all test run steps.

Table 8: Flash zone temperatures

STEP	1	2	3	4	5	6	7
REDUCED CRUDE (m^3/d)	BASE	BASE	BASE	BASE	+ 300	+ 300	+ 300
LVGO HOT REFLUX (m^3/d)	0	350	700	1000	0	450	1000
Temp Thermocouple 1 (°C)	383	384	381	381	381	381	381
Temp Thermocouple 2 (°C)	382	382	379	378	379	379	378
Temp Thermocouple 3 (°C)	380	381	378	377	377	377	378
Temp Thermocouple 4 (°C)	384	385	383	382	382	382	382
Max ΔT (°C)	4	4	5	5	4	5	4

LVGO Heat Transfer

As shown in Table 9, to increase the vapor rate to the LVGO section, and also to condense more liquid in this region, it is necessary to reduce the HVGO duty, with the consequent increase in the LVGO duty.

Table 9: Pumparound flow rates and duties

STEP	1	2	3	4	5	6	7
REDUCED CRUDE (m^3/d)	BASE	BASE	BASE	BASE	+ 300	+ 300	+ 300
LVGO PUMPAROUND (m^3/d)	BASE	BASE	+ 300	+ 600	+ 200	+ 200	+ 600
LVGO HOT REFLUX (m^3/d)	0	350	700	1000	0	450	1000
HVGO PUMPAROUND (m^3/d)	BASE	- 300	- 600	- 700	+ 600	BASE	- 700
LVGO DUTY (%)	BASE	+ 8	+ 25	+ 35	+ 11	+ 24	+ 42
HVGO DUTY (%)	BASE	- 1	- 8	- 6	+ 15	+ 7	+ 1

HVGO X LVGO Fractionation

It was not possible to notice any improvement in LVGO quality when the hot reflux was increased, as presented in Table 10.

Table 10: LVGO quality

STEP	1	2	3	4	5	6	7
LVGO HOT REFLUX (m^3/d)	0	350	700	1000	0	450	1000
LVGO spgr 20/4°C	0,9240	0,9235	0,9248	0,9245	0,9255	0,9239	0,9240
HVGO spgr 20/4°C	0,9450	0,9449	0,9441	0,9433	0,9433	0,9431	0,9427
LVGO 90% distillation (°C)	456	451	457	466	474	466	466
HVGO 10% distillation (°C)	418	416	411	415	413	412	410
GAP 10-90% (°C)	-38	-35	-46	-51	-61	-54	-56
LVGO 95% distillation (°C)	484	476	482	498	512	499	497
HVGO 5 % distillation (°C)	398	397	389	395	393	393	392
GAP 5-95% (°C)	-86	-79	-93	-103	-119	-106	-105

Conclusions

The REFAP vacuum tower revamp was very successful.

It was proven in a real operation that the empty spray section works very well for heat transfer purposes. Although for fractionation, it is still necessary to have packing in order to improve the vapor-liquid contact.

As a consequence of a sharp decrease in the column pressure drop, it was possible to reduce drastically the furnace outlet temperature by 30°C.

Previous studies and conclusions were reinforced in these present tests:

- The heat transfer in the LVGO section occurs at the bottom of the section;
- When increasing vapor load, it is necessary a higher empty space to condense more product;
- When increasing vapor load, which means C-Factor, the heat transfer is more difficult and there is a limit to this value.

In this work it was also proven the very good help that radius temperature profile gives us to analyze the quality of distribution. It was also observed that there is very little difference between the skin and internal temperature values.

References

Barbabela, B.A.; Guimarães, F.M.Q.G.; Waintraub, S.; Torres, G.A. Optimization of Heat Transfer Zones in Distillation Columns, Presented at X International Phoenics Users Conference, Melbourne, Australia, May 2004.

Waintraub, S.; Torres, G.A ; Serpa, A.L.W; Waschburger, R.L.; Oliveira, A.G. Removing Packings from Heat Transfer Sections of Vacuum Towers. Presented at the American Institute of Chemical Engineers 2005 Spring National Meeting, Atlanta, GA, April 2005.

Waintraub, S.; Torres, G.A ; Guimarães, F.M.Q.G.; Barbabela, B.A. Is it Necessary to Use Internals in Heat Transfer Sections of Refinery Columns? Presented at the American Institute of Chemical Engineers Annual Meeting, San Francisco, CA, November 2003.

26g DEA Treater Revamp Targets Foaming

Waldo E. De Villiers, Shell Global Solutions (US) Inc., 3333 Highway 6 South, Houston, TX 77082, William L. Green, Technical Services, Shell Chemicals LP, 15536 River Road, Norco, LA 70079, and Glenn Shiveler, Chemical-Gas Group, Sulzer Chemtech USA, 4019 South Jackson, Tulsa, OK 74145.

A DEA treater processing RCCU dry gas experienced infrequent but unpredictable loading events that negatively affected treating efficiency and capacity. A root cause analysis identified several equipment and operational aspects that could be improved. This paper will describe the equipment improvements that were implemented, including improved knock-out facilities, column internals and other design practices. The changes have proved to be effective with no significant loading events since startup. Performance of the tower before and after the revamp will be compared.

48a On Distillation Tray Weir Loadings

Mike Resetarits and A. Y. Ogundeji. Fractionation Research, Inc., 424 S. Squires St., Suite 200, Stillwater, OK 74074

Distillation tray liquid rates are usually described via a term known as "weir loading." A weir loading on a tray is simply the volumetric liquid flow rate per unit length of outlet weir. The last 50 years of distillation tray literature proposes to tray designers a very wide, and confusing, range of "weir loading criteria." It is often unclear as to whether the high-end criteria are referring to the points at which flooding occurs , or, the points at which the number of flow passes need to be increased. There are also low-end criteria which refer to spray fluidization points or points at which picket fence weirs need to be used. This paper explains the utility of the weir loading variable and the associated literature criteria.

TRAY HYDRAULIC OPERATING REGIMES AND SELECTIVITY

Ralph H. Weiland
Optimized Gas Treating, Inc.
Clarita, OK

ABSTRACT

Most gas treating contactors are highly liquid loaded, with good designs operating at vapor loads approaching the jet-flood limit. An optimal design also will have downcomers sized to approach the choke flood limit at the design point because this allows for the maximum tray active area, hence the smallest column diameter for a given throughput. At high weir liquid loads, trays operate predominantly in the froth regime where the gas dispersion rises through continuous liquid, interfacial areas are large, and the liquid is highly agitated. This results in high absorption efficiency for both CO_2 and H_2S. Low weir loads correspond to the spray regime where the liquid is dispersed as rather large droplets that are projected through a continuous gas phase as a spray. The droplets behave like nearly stagnant, rigid spheres and they have extremely small liquid-side mass transfer coefficients. Consequently, CO_2 absorption is severely retarded because its rate is controlled by the liquid-side resistance to mass transfer. This provides a concomitant improvement in H_2S removal.

A number of cases have come to light in which seemingly impossibly-low H_2S leak rates have been realized and much higher than normal CO_2 rejection rates observed from trayed columns. In every case, liquid rates were low and the weirs were long enough to have weir loads of 10 gpm/ft or less, putting the trays well into the spray regime. Simulations could be matched to plant performance data only by (1) using greatly reduced liquid-film coefficients for mass transfer (stagnant liquid), and (2) increasing the gas-side coefficient as expected for turbulent gas flow past drops.

The spray regime is quite effective in achieving ultra low H_2S leaks and greatly improved CO_2 rejection rates, and if conditions cause an existing plant to operate there, selectivity will probably greatly exceed what normally might be anticipated. Therefore, if processing goals and certain other circumstances are right, consideration should be given to operating in this region.

Introduction

Usually, contactors containing trays are initially designed to operate at 70 to 80% of jet and downcomer flooding rates using very conventional valve-type trays with few or no hydraulic performance enhancers such as tapered downcomers and swept-back weirs. These are safe designs because they use old and tried technology and leave plenty of room for easy capacity improvement should they fall short in the field or to accommodate the inevitable, future increases in plant throughput. Most contactors in gas treating applications have relatively high liquid to gas (L/G) ratios, usually quite a bit higher than in distillation, because the large amounts of acid gases to be removed require large solvent flows for absorption. A well-designed tray has a turndown ratio of at least 4:1; in other words, it can operate satisfactorily from a hydraulic standpoint down to about one

---

 Paper presented at the 2009 Spring AIChE Meeting, Tampa, Florida, April 26 – 30, 2009.

quarter or less of design rates. Below about a four-to-one turndown, liquid starts to weep through the trays to an extent that mass transfer performance might begin to suffer. At the liquid to gas ratios common in gas treating, trays normally operate in the froth regime. The biphase on the tray has its gas dispersed within a more or less continuous liquid, and the mixture flows across the tray as an intensely-agitated froth.

Sometimes a raw gas contains either very small amounts of H_2S and CO_2, or perhaps a small concentration of H_2S and a larger fraction CO_2, but the H_2S must still be removed to meet a parts-per-million specification. In such cases, only very modest solvent flow is needed to treat the gas with N-methyldiethanolamine, MDEA, (an amine that is fairly selective towards H_2S) so the liquid to gas ratio in the contactor is low. As the liquid flow is reduced, the froth gives way to a spray because there is insufficient liquid flow to maintain a reasonable volume of liquid on the tray. The still large vapor flow shatters the liquid, converts it to droplets and ejects it into the vapor space, and eventually the downcomer, not as a stream of continuous liquid flowing over a weir, but as a swarm of drops—a spray—thrown over the weir. What is important to realize is that the phase hydraulics, the interfacial areas, and therefore gas- and liquid-side mass transfer coefficients, can be radically different in the froth and spray regimes, resulting in the likelihood that selectivity will be significantly affected[1].

Tray Hydraulics

Maximum tray hydraulic capacity can be limited by any one of three types of flooding: (a) jet or entrainment flood, (b) downcomer backup flood, and (c) downcomer choke flood. The ability of a tray to operate at the lower end of the loading range is limited by (d) weeping and also by (e) entrainment that occurs even in the spray regime of operation as the liquid flow on the tray is decreased while maintaining a high vapor flow rate. The other low liquid rate limitation is vapor bypassing up downcomers as lowered liquid depths unseal the downcomers. In a perfectly balanced design, jet flood and choke flood will occur simultaneously and the column's diameter will be sized to be slightly more than adequate for the proposed loads; certainly pressure drop will not be high enough nor downcomer clearance tight enough to induce backup flood problems, nor will vapor bypass up downcomers.

Entrainment Flood

Figure 1 is a typical tray operating diagram showing stable limits. The portion of the upper operating line labeled Entrainment Flooding is in the Froth Regime and it corresponds to the upper limit of vapor flow where the tray begins to flood, the so-called jet flood line. This should not be interpreted as a finely drawn, sharp line because the definition of jet flood varies from vendor to vendor. Some vendors call 85% jet flood the vapor velocity at which 10% of the liquid is entrained from the tray and flows with the vapor to the tray above. Others call it the maximum useful capacity. Whether the definition is 80% or 90% jet flood corresponding to 10% or 20% entrainment is arbitrary and the pinpointing of maximum useful capacity is certainly subjective. However, near the jet flood limit the entrainment rate is exponentially sensitive to small changes in vapor rate and only a small increase in vapor flow will cause the column to flood. This is not a condition at which one would like to operate because the column approaches instability and becomes harder to control.

[1] Selectivity for H_2S over CO_2 is the direct result of the facts that (1) H_2S and CO_2 absorption rates are respectively controlled by vapor-side and liquid-side resistances to mass transfer and (2) the extreme slowness of the reaction of MDEA with CO_2 which leaves the CO_2 absorption rate almost completely unenhanced by reaction. Mass transfer rates, of course, are directly dependent on the mass transfer coefficients that pertain to each transferring species.

Vapor load is best described by the C-factor, usually based on the superficial vapor velocity, u_S, through the empty column but sometimes based on the gas velocity through the tray active area or even through the area of the actual openings (valves, etc.). The superficial-velocity definition is

$$C_S = u_S \sqrt{\frac{\rho_V}{\rho_L - \rho_V}}$$

A conventional valve tray with 13% straight downcomers operating at normal weir loads (50 – 150 gpm/ft) will flood when C_S is much above 0.38 ft/s. A more-modern high-capacity tray uses special downcomer treatments to recover much (and in the case of truncated downcomers, all) the area under the inlet downcomer, so it enjoys between a 10% and 20% higher capacity. This higher jet flood capacity is achieved simply by converting more of the tower area into tray active area.

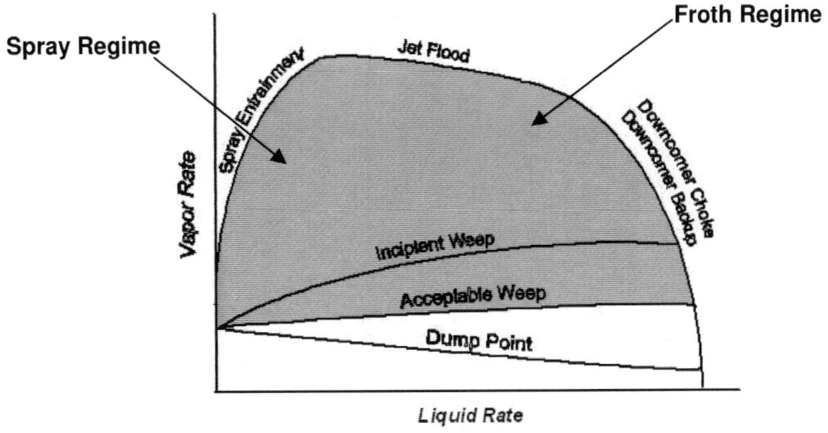

Figure 1 Tray Operating Diagram (adapted from Kister, 1992, p. 269)

One jet flood mechanism is froth entrainment by gas which carries liquid to the tray above in large enough quantities that downcomer capacity is exceeded on the top tray[2]. However, froths on operating trays are not well-behaved flows of uniform depth—massive eruptions and geysers occur with high frequency, with the liquid being thrown against the bottom of the overlying tray and forced through its perforations. At close tray spacing the froth may occupy the whole space between the trays: flooding occurs when the froth tries to expand even further. This may not be limited to close tray spacing—it is hard to make more than a slight distinction between froth occupying the whole volume between trays and numerous geysers throwing liquid at the upper tray. Of course, when foaming occurs, the whole jet (and downcomer) flood situation is exacerbated in the extreme.

[2] Except for the bottom-most and top-most trays, each tray in the column receives as much entrainment from the tray below as it loses to the tray above so it experiences very little change in liquid load caused by entrainment. The bottom tray actually has a lower liquid load than any other tray in the column because it receives no entrainment from below but loses liquid to the tray above. However, the top tray has most of its entrainment returned to it by the mist pads commonly installed in the top of the tower so the liquid load on the top tray is incremented by the entrainment rate from the tray below. Thus, the top tray has the highest liquid load and would be expected to flood first.

Downcomer Flood

Downcomers can flood by two possible mechanisms: excessive liquid backup inside the downcomer (backup flood), and choke at the downcomer mouth. The right-hand-side of Figure 1 does not distinguish between the two mechanisms, and indeed, it is often hard to distinguish between them even visually in a large-scale, well-instrumented pilot plant.

Backup Flood

As shown in Figure 2(a), the depth of liquid (or froth) in a downcomer is determined by (a) liquid depth on the tray, (b) the pressure drop across the tray and (c) the head loss for flow of aerated liquid under the downcomer skirt. The third factor can be controlled by using the right downcomer escape area but the second depends on tray pressure drop which is a function of the tray bubbling area, the type of tray opening being used, and the fractional hole area. Backup flood is the easiest to avoid.

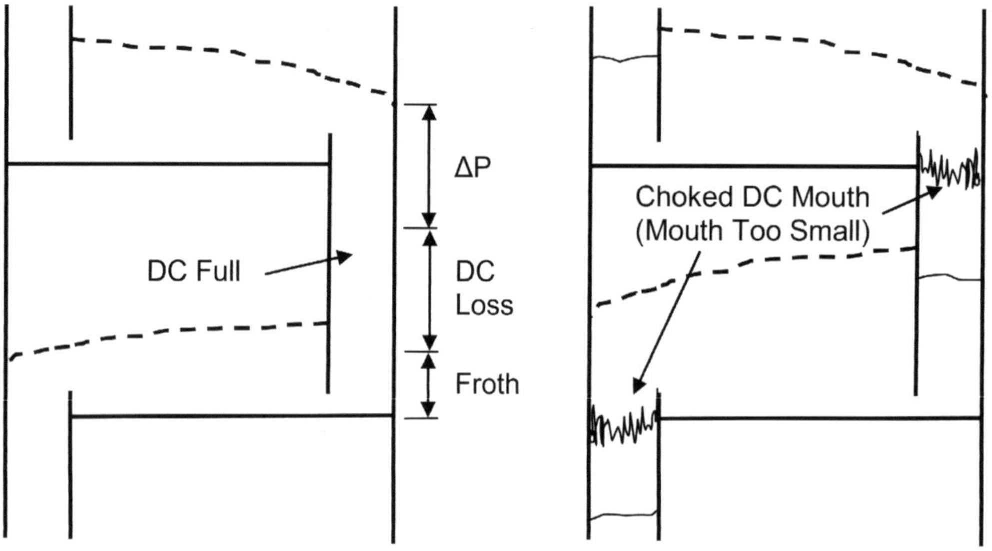

(a) Backup Flood **(b) Choke Flood**

Figure 2 Two Types of Downcomer Flooding (adapted from Kister, 1992, p. 272)

Choke Flood

The cross-sectional area (and to a lesser extent the shape) of the downcomer mouth is the primary factor that determines its capacity for accepting a given flow of liquid or froth. Referring to Figure 2(b), at choke, liquid does not crest over the weir. Instead, the downcomer is simply presented with a higher volume flow of froth than can be driven into it when hydrostatic head is at its greatest, equal to the tray spacing (the maximum head possible). Once the downcomer mouth becomes covered with froth, liquid height over the weir is no longer proportional to volume flow to the 2/3^{rds} power as in the Francis weir formula. Instead, liquid height over the weir becomes proportional to liquid rate raised to the 1.5 power; thus, any further increase in liquid flow requires *greatly* increased head to drive the flow into the mouth. Once the downcomer mouth becomes buried in froth, only small increases in liquid load are enough to throw the column into choke flood—the column becomes increasingly unstable and eventually impossible to operate[3].

[3] During a foaming event, perhaps downcomer choke flood might be considered the culprit because the froth volume is much increased, with the downcomer being asked to admit a greatly increased volume flow of even lower density material. Under choke conditions even a large increase in froth depth drives only a modestly increased flow. But this is

Weeping Through Tray Openings

Referring again to Figure 1, as the gas rate is decreased at a fixed liquid rate, a point is reached at which the tray starts to weep. This is the incipient weep point where the first droplets of liquid find their way downwards through a few perforations. Further decreases to vapor rate result in increased weep rates until the weep rate becomes, in some sense, excessive. Perhaps surprisingly to distillation experts, there have been instances where during a turnaround all the trays in a conventional 20-tray contactor have been found in a heap in the tower sump, yet the column was making on-specification gas (usually total acid gas removal) at design rates just before the turnaround! Obviously what defines a weep rate as excessive depends very much on the operation. However, when slipping CO_2 is important or meeting *both* an H_2S *and* a CO_2 specification is the objective, the column must be more tightly designed. Too much weeping compromises the separation because to some extent it spoils the split between CO_2 and H_2S.

Spray Regime

Kister and Haas (1987) have shown that entrainment rate goes through a minimum with weir liquid load (Figure 3). The minimum might be considered to be somewhere in the transition between the froth and spray regimes, the spray regime lying to the left. However, the transition itself is very gradual so there is no "point" associated with it. The spray regime is entered at higher liquid rates in hydrocarbon systems (cyclohexane–n-heptane in the Figure) than in air-water (much closer to amine treating fluids). The higher the gas rate, the higher the weir load at transition; however, unless the tray is operating at an extremely low vapor rate, transition to the spray regime occurs at a weir liquid load (volume liquid flow per lineal length of weir) between 15 and 30 gpm/ft and the entrainment rate climbs super exponentially as liquid load is further reduced.

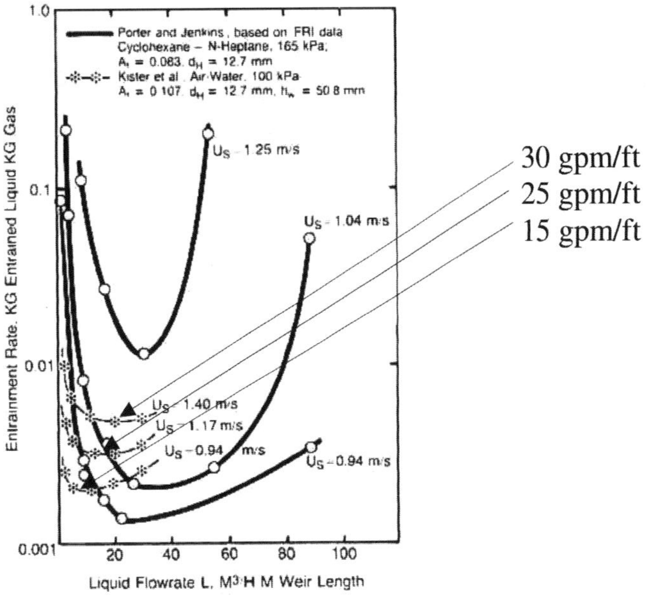

Figure 3 Dependence of Entrainment Rate on Weir Load at Various Gas Rates for Cyclohexane – n-Heptane and Air–Water (Kister & Haas, 1987)

---

speculation because once flooding begins, it is extremely hard to ascribe one mechanism or another as the cause—they are all connected.

In terms of mass transfer, the gas-liquid interfacial area per unit volume of spray can be expected to be smaller than the same area in good-quality, deep froth. The droplets, being small and probably containing some amount of contamination as well, exhibit the internal hydraulics of essentially rigid spheres. This means that the transfer of acid gases into the interior of the drops must take place by purely molecular diffusion, a very slow process compared with transfer into the highly-agitated, turbulent liquid in froth. Also, as noted by Lockett (p. 39, 1986), photographic evidence indicates that drop breakup and coalescence does not occur during flight so even this method of getting dissolved gas into the drop's interior is not available. Thus, the liquid-side mass transfer coefficient might be expected to be *very* much smaller than in a froth. Because CO_2 absorption is controlled largely by resistance in the liquid phase, the liquid-side mass transfer coefficient for CO_2 absorption, and therefore the CO_2 absorption rate itself, should be seriously reduced, i.e., a great deal more CO_2 should be rejected and CO_2 slip significantly increased.

It is perhaps less easy to speculate on the effect of spray-regime operation on gas-side mass transfer resistance. In surface-tension-negative systems (where surface tension decreases as liquid flows down the column), higher distillation efficiencies are observed in the spray regime (distillation is generally controlled by vapor phase resistance) because smaller drops are produced. However, gas treating is a surface-tension-neutral system. So surface tension effects probably do not pertain. However, highly turbulent gas flow past drops is markedly different from gas bubbles and jets sparging into and rising through liquid. Unconfined gas flow past large drops probably is much more turbulent than gas in froth, especially near the interface, so higher gas-side mass transfer coefficients might be anticipated. Flow past spheres transitions from laminar to turbulent at a Reynolds number of about unity. At atmospheric pressure, a 1-mm drop (typical of a spray, Kister, 1992, p. 292) has a terminal velocity of about 7 m/s and falls at a Reynolds number of about 500, i.e., in the fully-turbulent region. It is not hard to imagine that fully turbulent flow past a myriad of drops results in large values for the gas-side mass transfer coefficient since molecular diffusion through the gas plays a role only quite close to the drop surface where concentration gradients are unavoidably very steep.

The point is that the spray and froth regimes present completely different types of fluid flow and, because hydraulics is one, if not the, major determining factor for mass transfer, the mass transfer performance in these two regimes ought to be very different. The effect of the difference is highly magnified in selective gas treating. Does this really lead to differences in CO_2 slip and, therefore, H_2S absorption? The two sets of plant performance data presented next suggest that indeed it does.[4]

Simulation and Field Data

Before embarking on using the ProTreat™ simulator's mass transfer rate-based column model to explain why certain treating plants with absorbers operating in the spray regime have been observed in the field to perform so much better than they "should", it is important to be convinced that simulation is capable of truly *predicting* plant performance at all.

---

[4] It should be noted that operating in the spray regime can result in reduced tray vapor-handling capacity unless the tray is carefully designed. There are ways to regain froth regime operation if this is desired. One is to reduce the number of tray passes—one of the main reasons for using multiple tray passes in the first place is to *reduce* excessive weir loads but, if the weir loads are already too small, using multiple passes is counterproductive and fewer passes should be used. The other approach is to use so-called weir blocks, or picketed weirs, which effectively shortens the weir length by placing a series of vertical barriers (pickets) on the weirs to force part of the liquid spray back onto the tray and make the effective weir length shorter.

™ ProTreat is a trademark of Optimized Gas Treating, Inc. Other trademarks are the property of their owners.

Predicting Plant Performance

A mass transfer rate-based column model uses the fundamental *mass transfer* characteristics of the specific tower internals (hydraulic- and property-dependent, phase, mass transfer coefficients of trays, random packing, and structured packing) together with phase equilibrium and reaction kinetics to calculate the actual mass transfer *rates* of individual species occurring between the liquid and vapor phases flowing on real trays and through packed beds. The approach is completely analogous to the way heat transfer calculations have been done and heat exchangers have been designed for nearly a century. The real difference is the transferring entities: heat in one case and chemical species in the other, so temperature differences are used in one case and concentration differences in the other. Where heat exchangers' heat transfer properties are correlated, for example, in terms of shell and tube-side heat transfer coefficients as functions of exchanger hydraulics and the transport properties of the heated and cooled streams, the mass transfer properties of trays and packing are correlated as liquid-side and vapor-side mass transfer coefficients, again as functions of hydraulics and the transport properties of the two phases. The analogy is strong. Phase equilibrium and the number of species make mass transfer calculations a lot more challenging than heat transfer but this is only a computational technicality—if the underlying correlations themselves are reliable, heat-transfer-rate and mass-transfer-rate models predict equipment performance with equal trustworthiness and fidelity. *Engineers don't use efficiency-corrected equilibrium heat exchanger models, and there is no longer necessary to use their mass transfer equivalent in tower design.* This is especially the case in amine treating where different phases dominate the absorption rates of each component, the components have very strong thermodynamic interactions via chemical reactions, and the component-dependent tray efficiencies typically range from a few percent to 20%.

The mass transfer rate-based approach to column simulation as implemented for example in Aspen Plus® and Pro/II™ under the name RATEFRAC® (a general purpose module), by several solvent vendors, and within ProTreat™ (specifically for gas treating), has been used to simulate hundreds of columns around the world with outstanding success. The approach has been in continuous commercial use for some 20 years, so there can be little doubt that it is a sound one. The advantage of mass transfer rate-based simulation is that it is truly and reliably *predictive*, without leaning even slightly on any knowledge of plant performance to tweak the model in order to get the right results. Today, the mass transfer simulation of columns is on just as solid a footing as the heat transfer simulation of heat exchangers and the calculation of pressure drop through piping networks.

Case Studies

Both case studies involve plants that are treating raw gas having parts per million H_2S, in the first case 60 ppmv and in the second 240 ppmv. Both plants produce treated gas having about 1 ppmv H_2S and, because neither is treating gas containing a lot of CO_2, only low solvent rates are necessary to process the gas. In other words, the contactors in these plants are operating with unusually low L/G ratios.

Case Study 1

INEOS* GAS/SPEC* SS-3 solvent at 45 wt% strength is used to treat gas containing 60 ppmv H_2S and 0.94 mol% CO_2 at about 460 psig. Figure 4 shows the flowsheet.

---

* INEOS and GAS/SPEC are trademarks of INEOS Oxide, a division of INEOS Americas, LLC

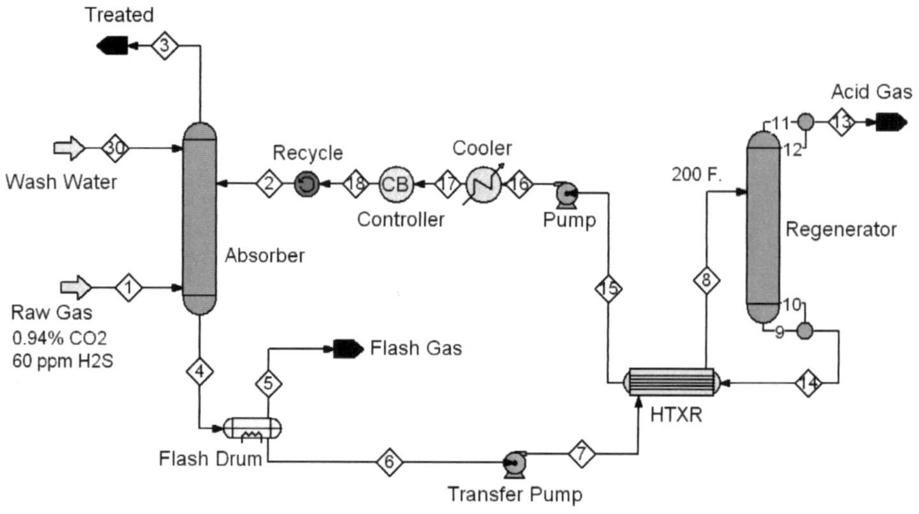

Figure 4 PFD for Case Study 1

Straightforward simulation of this process predicts that the treated gas should contain 27 ppmv H_2S and the CO_2 level should have been reduced to 0.636 mol% with the rich amine loaded to 0.634 moles of CO_2 per mole of SS-3 solvent. However, a series of carefully planned and executed plant trials showed unequivocally that the raw gas, varying between 50 and 70 ppmv H_2S was actually treated to between 1.0 and 1.2 ppmv H_2S and the treated gas was consistently at 0.865 mol% CO_2 (93% CO_2 slip!) with a rich solution CO_2 loading of only around 0.13 mol/mol. The simulation results weren't even close to reality!

These are 1-pass valve-type trays and because of the low L/G ratio they have only 5% downcomers, i.e., 90% active area. The trays run at about 50% of jet flood with the downcomers at only 11% of choke flood. Jet flood and choke flood are not issues. The striking hydraulic parameter, however, is the weir load—only 11 gpm/ft. At a superficial gas velocity of 1.27 ft/s, this puts the trays into the spray regime of operation; on the other hand, the simulation is based on mass transfer characteristics that pertain to the froth regime[5]. The measured plant data are accurately reproduced by the simulation if the liquid-side and gas-side coefficients are decreased and increased, respectively, by a factor of 10. The simulated results in this particular case were not very sensitive to reasonable changes in interfacial area. It is interesting to note that decreasing the liquid-side coefficient alone produced the observed CO_2 slip and also resulted in a significant (albeit insufficient) improvement in H_2S pickup because less solvent was used up by over-absorption of CO_2. A further significant increase in gas-side mass transfer coefficient was needed to reach the observed H_2S treat.

Case Study 2

For this case, only data for the contactor were available. A lower pressure gas (155 psig) was being treated than in Case 1, using a solvent mixture comprising 19.15 wt% MDEA and 10.85 wt% DEA (30 wt% total amine) flowing at only 5 gpm to the contactor. The raw gas contained 2.1

---

[5] Much work has been reported on the froth-to-spray *transition*. However, no work appears ever to have been done, and none has been reported, which measures such fundamental mass transfer parameters as film coefficients and interfacial areas for mass transfer in the spray regime, possibly because the general recommendation is not to operate there. Therefore, *all mass transfer rate*-based simulators do their calculations using mass transfer characteristics that pertain to froths, and this takes care of the overwhelming majority of operating distillation and absorption columns.

mol% CO_2 and 240 ppmv H_2S. Lean amine CO_2 and H_2S mole loadings were 0.0136 and 0.0005, respectively. The H_2S leak from the absorber was measured to be 1 ppmv and the treated gas contained about 1 mol% CO_2. However, simulation based on froth-regime, mass-transfer coefficients predicted 23 ppmv H_2S and 0.72 mol% CO2. Again, too much CO_2 pickup is predicted and, partly as a consequence, simulated H_2S removal is too low by a very substantial margin. The hydraulic calculations gave a weir liquid load equal to the amazingly-low value of 2.8 gpm/ft, making the spray regime an absolute certainty. There were no special weir treatments. But, just as in Case 1, if the liquid-side and gas-side coefficients are decreased and increased, respectively, by a factor of 10, the plant data are perfectly matched by the simulation. It should be mentioned that the trays in this example were turned down to only 11% of jet flood so serious weeping was possibly another issue; nevertheless, regardless of whether there was weeping or not, the liquid was definitely in the form of a spray and spray regime mass transfer corrections should apply. The results indicate that indeed they do.

Factors of 10 changes in individual phase mass transfer coefficients are certainly not small. However, it should be realized that the flows of the phases change from turbulent to stagnant in the case of the liquid, and from *relatively* quiet for gas bubbles to very turbulent near drop surfaces in the case of the gas. The size of the corrections emphasizes the radical difference in the flows.

Discussion

For the most part, gas treating engineers seem unaware of the important distinction between spray and froth regimes of tray operation, and the effect of weir load in determining where a given set of trays actually operates. More importantly, the impact of froth versus spray flow regime on *selective* performance appears never before to have been addressed quantitatively. However, the idea that a laminar or stagnant liquid will promote CO_2 rejection is certainly not new. Sigmund and Butwell (1981) for example developed a special tray intended to produce a laminar liquid flow with the specific objective of increasing CO_2 slip and, therefore, improving H_2S removal rates. Their idea was to produce laminar liquid jets on trays, with the liquid flowing into truncated downcomers with perforated bottoms. Although a version of their tray design was installed in a New Mexico gas plant owned by El Paso Natural Gas, no serious effort seems to have been made to commercialize this patented tray.

The *normal* weir loading range for conventional trays is from roughly 40 or 50 gpm/ft to about 150 gpm/ft. Modern high-performance trays have superior capacity because they trade downcomer area for active area but there is no reason to believe maximum operable weir loads are any higher than a conventional tray. In gas treating applications L/G ratios are generally high because most gases contain high concentrations of acidic components and the solvents have limited capacity, so high solvent rates are required relative to the gas flows being treated. Furthermore, except in tail gas treating and more recently CO_2 capture from power plant flue gases, treating pressures are usually in the 100s of psi range. These factors make spray regime operation rare in gas treating, perhaps explaining the general lack of appreciation for the very existence of this regime.

The two cases discussed here presented quite a puzzle when they first came to light. It just seemed physically impossible to produce 1 ppmv H_2S gases under the stated conditions and slip as much CO_2 as was claimed. No amount of supposition about such things as possible effects of solvent contamination and tray damage made any sense. Perhaps a gamma scan might have led to earlier resolution; however, in both cases the contactors were treating the gases perfectly satisfactorily so there was certainly no commercial incentive to spend money diagnosing a non-problem. The inquiry was driven by the failure of the simulations to come even close to measured performance. It was intriguing that simulation of these two plants, operated by different companies and using quite different solvents, showed such closely similar deviations from measured performance.

When more detailed tray hydraulic calculations were done, it became blatantly evident that in both cases the contactors were operating well into the spray regime[6]. Therefore, the basic mass transfer coefficients used in the simulations were almost certainly at fault. Confirmation of this thinking was that the simulations required virtually identical adjustments of basic mass transfer coefficients to bring the results into close conformity with measurements. One of the outcomes is that much more detailed tray hydraulic calculations are now done as a routine part of ProTreat simulations.

Very low L/G ratios are rare in gas treating. However, in instances where they exist, much increased CO_2 rejection and far better H_2S removal seem to result. So, should spray regime operation be encouraged (for example, by using multi-pass trays with long weir lengths when a one-pass tray is sufficient) as a means to promote better selectivity in gas treating? Because of the normally high L/G ratios common in gas treating, there is only infrequent opportunity to consider such an approach, at least using conventional trays. Furthermore, conventional "wisdom" is to stay away from this region[7], supposedly because entrainment rates are higher, so superficial gas velocity would have be reduced (tower capacity sacrificed), necessitating a more-expensive, larger-diameter tower. However, be this as it may, there are circumstances such as the ones in the Case Studies reported here, where a very significant advantage can be had by purposefully operating in the spray regime. Indeed, two of the world's highest capacity (and most expensive) trays, Shell's ConSep™ tray and the Koch-Glitsch ULTRAFRAC® tray are in a sense co-current flow trays by design, and both operate primarily in the spray regime over their entire operating range. These trays might be attractive for obtaining selectivities in a range well beyond the reach of conventional cross-flow trays although their cost effectiveness would certainly be application dependent. Nevertheless, using quite conventional trays, some selective contactors have been enjoying the great selectivity benefits of the spray operating regime for years—what is new is knowing the reason for their success. Understanding certainly does not increase risk—it decreases it—so where possible, there is a real incentive to operate in the spray regime because selectivity is tremendously improved.

Although trays are often not the best choice of tower internals for very low L/G operations, in the present case, simulation and experience indicate that when it comes to selectivity neither random nor structured packing can come even close to trays operating in the spray regime.

Literature Cited

Kister, H. Z., *Distillation Design*, McGraw-Hill, Inc., New York, NY, 1992.

Kister, H. Z., and J. R. Haas, *I. Chem. E, Symp. Ser.*, 104, p. A483 (1987).

Lockett, M. J., *Distillation Tray Fundamentals*, Cambridge University Press, 1986.

Sigmund, P. W., and K. F. Butwell, US Patent 4,278,621, July 14, 1981.

---

[6] The trays used in these studies were not designed specifically for extremely low weir loads and spray regime operation. However, they used inlet sumps so vapor bypassing up the downcomers was not likely an issue.

[7] The author thanks Henry Z. Kister for pointing out just how ill-founded this conventional wisdom really is. In fact, trays can have even higher capacity in the spray than in the froth regime *if they are properly designed.* Obviously, downcomers must be designed to remain sealed without reliance on the outlet weir; an obvious way to do this is to use a sufficiently deep inlet sump.

48c 4-Pass Tray Design Techniques

Daniel R. Summers, Sulzer, 4019 South Jackson Avenue, Tulsa, OK 74107

Four Pass (4-pass) trays have been employed for many decades in the chemical and refining industries. Many papers have been published on this subject but none give a concise methodology on how to design and balance the V/L ratio on each tray panel. This paper will discuss the methodology of balancing 4-pass trays regardless if the design is based on equal bubbling area or on equal flow path length and whether or not the trays have vapor tunnels. In addition, deficiencies in current pressure drop methods will be discussed and how they relate to and affect 4-Pass tray design.

48d Preventing Maldistribution in 4-Pass Trays

Richard W. Dionne, **Walter J. Stupin**, Matthew` Olsson, and Henry Z. Kister. Fluor Corporation, 3 Polaris Way, Aliso Viejo, CA 92698

When the liquid load is high, the tray liquid is often split into two or more flow passes. Perfect symmetry is readily achievable in two pass trays, where the design of one pass is identical to the other, but not so where a further number of passes is involved. The non-symmetry renders multi-pass trays prone to vapor and liquid maldistribution. Three-pass trays are completely non-symmetrical and therefore uncommon. Four-pass trays retain the half-tray symmetry and are common and popular in industrial practice, but are still prone to maldistribution. Vapor and liquid maldistribution on 4-pass trays can lead to major loss of efficiency and capacity.

Tools for reliably predicting this maldistribution are scarce. Most of the available tools are proprietary and unreliable. In the absence of adequate predictive tools, achieving adequate pass distribution and keeping the distribution within the recommended distribution ratio is elusive and often unreachable. We are seeing far too many recent designs in which the distribution ratios were far worse than the recommended. We have seen this maldistribution causing major loss of efficiency, capacity, or both, in four-pass trays.

Our paper presents the key equations and procedures to calculate the vapor and liquid distribution on four pass trays. It then tests our method against two case studies in which the liquid distribution is known from gamma scans. Although focused on four-pass trays, the principles are readily extensible to other number of passes.

48e De-Bottlenecking Splitters with SUPERFRAC® High-Performance Trays

Izak Nieuwoudt and Michael J. Lockett. Koch-Glitsch, 4411 E 37th Street N, Wichita, KS 67220

Significant capacity improvements and reduction in specific energy consumption in splitters can be realized by replacing conventional trays with high-capacity, high-efficiency, multi-pass SUPERFRAC® trays. Two industrial case studies will be covered in this presentation: (a) A C2 splitter that was de-bottlenecked to give higher throughput and higher purity ; (b) A C3 splitter that was de-bottlenecked to give higher throughput.

48f Upgrading a Super-Fractionator C4-Splitter with 4-Pass High Performance Trays

Giuseppe Mosca[1], Elena Tacchini[1], Stefan Sander[1], **Dirk Rottger**[2], Andreas Wolff[2], Armin Rix[3], and Gunnar Schilling[4]. (1) Technology, Sulzer Chemtech, Piazza Duca D'Aosta, 12, Milano, 20124, Italy, (2) Technology, Evonik Oxeno AG, Paul Bauman Str. 1, Marl, 45764, Germany, (3) Engineering Projektierung 1, Evonik Degussa AG, Paul Bauman Str. 1, Marl, 45764, Germany, (4) Technology, Evonik Degussa AG, Paul Bauman Str. 1, Marl, 45764, Germany

The column is located in an integrated C4-Plant at Evonik Degussa's site in Marl, Germany. Its key function is the sharp separation of light C4-olefins from heavier C4s. The column was originally equipped with a high number of conventional 4-pass valve trays, installed at a tray spacing as low as 300 mm. The throughput and especially the production rate of the desired distillate product were limited by excess carryover of heavier C4s into the distillate. Following growing demand, a capacity increase by higher feed flow rate to the column and / or higher overhead product recovery was planned. Distillate purity had to remain at established values and the revamp had to be performed during a planned 3 week plant shutdown.

The paper covers all essential steps of the revamp.

First, the problem definition via process simulation, column scan and detailed review of existing trays is outlined. Bottlenecks in the existing tray design were revealed.

In the second part, the design procedure for high capacity 4-pass truncated downcomer trays at low tray spacing is outlined.

The safe exchange of more than 160 metric tons of tray material during a three weeks plant shutdown was made possible by a combination of hardware designed for easy installation, extensive pre-assembly of tray elements before the shutdown and carefully planned installation work.

A comparison of column performance before and after the revamp is given. Before the revamp, the column was run at maximum capacity at most times, and frequent operator action was required to keep the distillate within specification. The new trays give a twofold advantage: hydraulic capacity has been increased by more than 15%, while the reflux ratio could be significantly reduced. This apparent improvement in tray efficiency results in a further capacity increase. The column now runs stable at 20% higher capacity, requires less operator action and the products are in specification.

48g Optimization of Multi-Pass Distillation Tray Column by CFD Method

Xueying Gao, **Chunjiang Liu**, Xi-Gang Yuan, and Guocong Yu. State Key Laboratory of Chemical Engineering, Tianjin University, 92 Weijin Road, Nankai District, Tianjin, 300072, China

Multi-pass distillation tray column has been widely used in chemical and petrochemcial industries. Traditional design method of multi-pass distillation tray column was mainly based on empirical models. There was very few published work on how to optimize multi-pass distillation tray by CFD method in order to increase the separation efficiency. In this paper, CFD method are employed to simulate the flow and mass transfer performance in a 12.6m in diameter four-pass distillation tray column and optimized configuration of the tray was proposed.

Based on our previous work (AIChE meeting, Distillation 2003: 479-496), a CFD model to simulate the flow and mass transfer performance of the four-pass distillation tray was proposed. Using this model, flow pattern and concentration distribution of the liquid on different trays of the column can be simulated. Using the concentration profile of the vapor and liquid phase on different trays, column efficiency can be calculated. By changing the configuration of the tray and downcomer, which include tray length, weir length, weir height, shape of the downcomer and shape of the inlet weir, flow patterns of the liquid phase on the tray can be modified. By analyzing the column efficiency at different flow patterns, optimized configuration of the tray and downcomer can be obtained.

At first only single tray was simulated. During the simulation, two important assumptions are made, one is the length of flow path between side-tray and mid-tray are equal, the other is initial velocity and concentration of vapor and liquid at the inlet are uniform. The results show the existence of liquid circulation area is apparent on side-tray rather than mid-tray, which is in agreement with literature. From the velocity and concentration profile it can be found that the maldistribution on side-tray is clearer than mid-tray due to existence of liquid circulation and velocity distribution has much influence on concentration distribution. Tray efficiency at different liquid load ratio can be calculated. It was found that tray efficiency is the highest at the liquid load ratio of 0.5 between side-tray and mid-tray.

Then the structure of downcomer was modified into a specified configuration, which can lead the liquid to flow along a certain path. Simulation result shows liquid circulation area becomes smaller and velocity and concentration distribution on side-tray become relatively uniform. In this case, mass transfer is intensified and tray efficiency was increased by 6.4%. It is concluded that the modified structure of tray was much better.

Keywords four-pass distillation tray CFD mass transfer tray efficiency tray optimization

64a Packed Tower Internals Important to Tower Efficiency

Frank Rukovena Jr., Fractionation Research, Inc., 117 W. 5th, Suite 410, Bartlesville,, OK 74003

The paper reviews the importance of liquid distribution to packed tower performance. Given the energy intensity of the distillation process it is essential that distillation towers operate efficiently. Miss-design or installation of tower internals can cause any where from minor to catastrophic efficiency loss.

The results of Fractionation Research, Inc's many years of research with various types of liquid misdistribution and its effects on packings efficiency under distillation conditions will be reviewed.

Many of the items to be considered when designing a liquid distributor will be presented. Things to watch out for when installing packing and internals will be discussed based upon the experience gained by FRI(SM) staff from hundreds of hardware installations performed installing packing and tray hardware in FRI's 1200mm diameter pilot plant towers.

Key Words: Internals, distillation, packing, liquid distribution,

64b On-Line Troubleshooting Techniques for Packed Distillation

Lowell Pless, Tracerco, 4106 New West Dr, Pasadena, TX 77507

Mass transfer efficiency of distillation or separation columns with packing is very sensitive to the quality of liquid/vapor distribution. Being able to confirm suspicions of liquid/vapor mal-distribution and identifying the cause while troubleshooting the operation of packed columns is extremely advantageous. The opportunity to accomplish this while the equipment is on-stream so plans for remediation can be made without panic is even more beneficial.

There is a wide range of on-line diagnostic services available that can provide critical data offering real-time information on how packed distillation towers are actually hydraulically operating. The most common application is gamma scanning. Gamma scans provide a density profile of the internal process of operating distillation columns and other process vessels. The density profile can be used to diagnose hydraulic operating conditions of mass transfer devices and determine issues such as damage to internals, flooding, degree of entrainment or weeping, liquid levels on trays and distributors, liquid distribution through packed beds, etc. The presenter will show case studies where scanning revealed vital process information that helped identify or solve an operating problem, or helped make a revamp/re-design successful on packed columns.

The CAT-Scan technique is a specialized horizontal gamma scan used to generate a topographic profile of the internal cross-sectional density of process equipment. This profile is useful for the detailed study of liquid flow distribution through packed columns. Applications involving tracers to study internal flow patterns will also be discussed.

A frank discussion of the benefits and limitations of these techniques will be presented, as well as areas where research and development is currently underway. As all past diagnostic technology innovations have started from a customer need, the presenter hopes to stimulate ideas from the audience that may result in the development of additional mass transfer diagnostic innovations.

1.0 Experimental Method

1.1. Flow Rates

This research sought to document the physics occurring in the vapor phase of a vapor-liquid contactor. However, it was necessary to use water as the process medium to allow for the creation of neutrally buoyant particles. Therefore, similarity theory was employed to choose water flow rates which would exhibit behavior analogous to that seen in the vapor phase of industrial contactors.

Flow rates in industrial vapor-liquid contactors are traditionally quantified in terms of an F-factor (units of velocity times the square root of density, denoted as F_S) and typical operating rates are from 1.8 – 2.5 m/s * $(kg/m^3)^{0.5}$ [1.5 – 2 ft/s * $(Lb_m/ft^3)^{0.5}$]. Using the definitions of F-factor and Reynolds number, one can set the water and air Reynolds numbers equal and solve for water velocity in terms of the corresponding vapor F-factor. This calculation is tabulated in Table 1. Data was taken in this study at flow rates of 4.54, 5.68, 6.81, and 7.95 m^3/hr [20, 25, 30, and 35 gpm].

Table 1 - Similarity theory calculation of flow rates in a 14.6 cm [5.75 in] ID column from F-factor.

Air		H_2O				
F_s		Superficial				N_{Re}
(m/s $[Kg/m^3]^{1/2}$)	(ft/s $[lb/ft^3]^{1/2}$)	m/s	ft/s	m^3/hr	GPM	unitelss
1.83	1.5	0.111	0.366	6.7	29.6	16,277
1.95	1.6	0.119	0.390	7.2	31.6	17,362
2.07	1.7	0.126	0.414	7.6	33.6	18,448
2.20	1.8	0.134	0.439	8.1	35.5	19,533
2.32	1.9	0.141	0.463	8.5	37.5	20,618
2.44	2.0	0.149	0.488	9.0	39.5	21,703

1.2. Particle Introduction/Recovery

The particles were introduced through a "T" in the process piping immediately after the flow meters. A 15.24 cm [6 in.] piece of Tygon tubing with valves on either end was connected to this "T" (**Error! Reference source not found.**). In order to introduce particles at steady state, both valves were left closed while the system was brought to a steady flow rate. Then, the outer valve was opened and particles placed in the tubing. Next, the outer valve was closed and then the inner valve was opened and the particles allowed to enter the process stream.

The particles were introduced in batches of five to ten particles at a time. This increased the odds of seeing and recording particles in each experimental run. After each experimental run, the pump was shut off and the filter isolated from the process. The filter could then be opened and the particles recovered.

1.3. Movie Acquisition

The X-ray scanner recorded movies on the computer system used to monitor and control the scanner. This computer system displayed radiographs in real time and read radiographs into its buffer at a 30 Hz sampling rate. At this rate, the buffer could accommodate three seconds of radiographs. Movies in DICOM format were created by dumping the buffer to the hard drive.

2.0 Results and Discussion

Thirteen experiments were successfully videotaped. These experiments were conducted at flow rates ranging from 4.54 to 7.95 m^3/hr [20 and 35 gpm]. Videos were taken of flow through three regions: the first (lowest) packing element, the joint between the first and second packing elements, and the second (middle) packing element. Flow within packing elements largely followed linear paths, while flow at the joints and wall exhibited chaotic flow with recirculation zones and discontinuous velocities.

Features of known length within the radiographs were used to create a conversion between image and real world dimensions. This, combined with the assumption that the 30 Hz frame rate equaled a one-thirtieth of a second exposure, enabled the calculation of instantaneous velocities using the initial and final positions of the particles in each radiograph. Two experiments recorded velocities through the lower joint under a 7.95 m^3/hr [35 gpm] volumetric flow rate. Figure 1 graphs all velocities recorded in these experiments as a function of height. On average, the particle velocities closely approximate the calculated interstitial velocity of 14.4 cm/s [5.67 in/s]. This suggests that the physics and velocities captured are representative of the bulk flow.

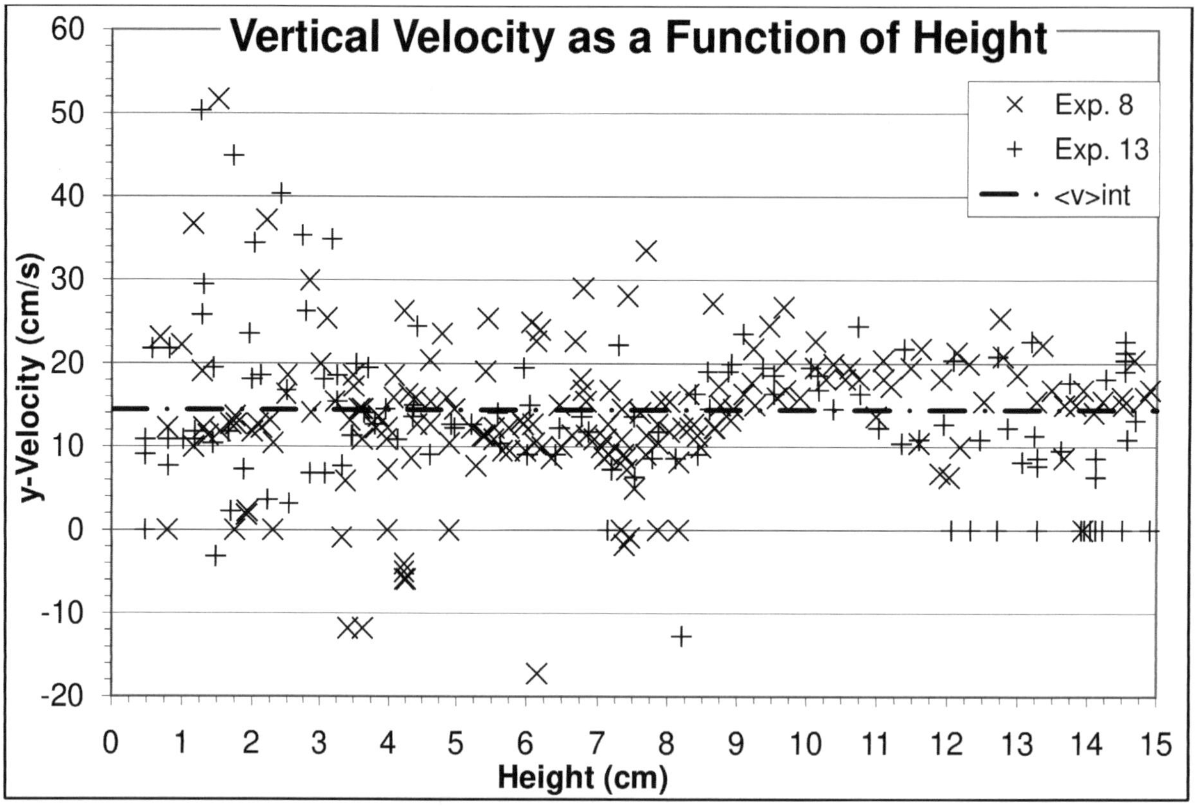

Figure 1: Vertical velocities measured at 7.95 m^3/hr [35 gpm] through the lower joint plotted as a function of their height in the image.

Table 2 shows an excerpt of velocity calculations performed on data taken from the fifth experiment. The displacement per frame is calculated directly from the first and last positions occupied by the particle. The instantaneous velocity of the experimental fluid can be calculated by multiplying the frame displacement by the frame rate employed. Because locations within the image are indexed by horizontal and vertical positions, component velocities are obtained directly. This is ideal for comparison with CFD predictions as many CFD software packages report local velocities in component notation and allow the user to specify custom coordinate systems to further simplify matching the predicted and measured velocities. Table 2 also shows the equivalent air velocity calculated using similarity theory.

Table 2: Velocity calculations performed for the fifth experiment.

	Positions (px)				Displacement (cm)		Component Velocity (cm/s)			
	Initial		Final				H_2O		Air (Equivalent)	
Frames	x	y	x	y	x	y	x	y	x	y
60	986	857	971	809	-0.227	0.726	-6.8	21.8	-102	326
61	966	813	942	713	-0.363	1.512	-10.9	45.4	-163	679
62	935	712	950	571	0.227	2.132	6.8	64.0	102	958
63	953	603	905	493	-0.726	1.663	-21.8	49.9	-326	747
64	912	503	846	449	-0.998	0.816	-29.9	24.5	-448	367
65	847	459	831	429	-0.242	0.454	-7.3	13.6	-109	204
66	842	438	797	432	-0.680	0.091	-20.4	2.7	-306	41

67	802	440	758	449	-0.665	-0.136	-20.0	-4.1	-299	-61
68	758	450	774	419	0.242	0.469	7.3	14.1	109	211
69	774	432	786	406	0.181	0.393	5.4	11.8	82	177
70	783	424	773	385	-0.151	0.590	-4.5	17.7	-68	265
71-73	Stationary						0.0	0.0	0	0
74	776	382	752	407	-0.363	-0.378	-10.9	-11.3	-163	-170
75	773	390	747	427	-0.393	-0.559	-11.8	-16.8	-177	-251
76	750	422	747	458	-0.045	-0.544	-1.4	-16.3	-20	-245
77	749	451	754	484	0.076	-0.499	2.3	-15.0	34	-224
78-84	Stationary						0.0	0.0	0	0

Figure 2 - Figure 6 show particle paths through the lowest packing element recorded during experiments with flow rates of 5.68 m^3/hr [25 gpm]. In each figure, the particles show a tendency to flow smoothly at an angle which corresponds to the corrugation angle of the packing (45°). Figure 3 shows the particle curve upward as it enters the packing and follow an increasingly linear path through the packing. This suggests the channels enforce a uniform and linear flow pattern within the packing.

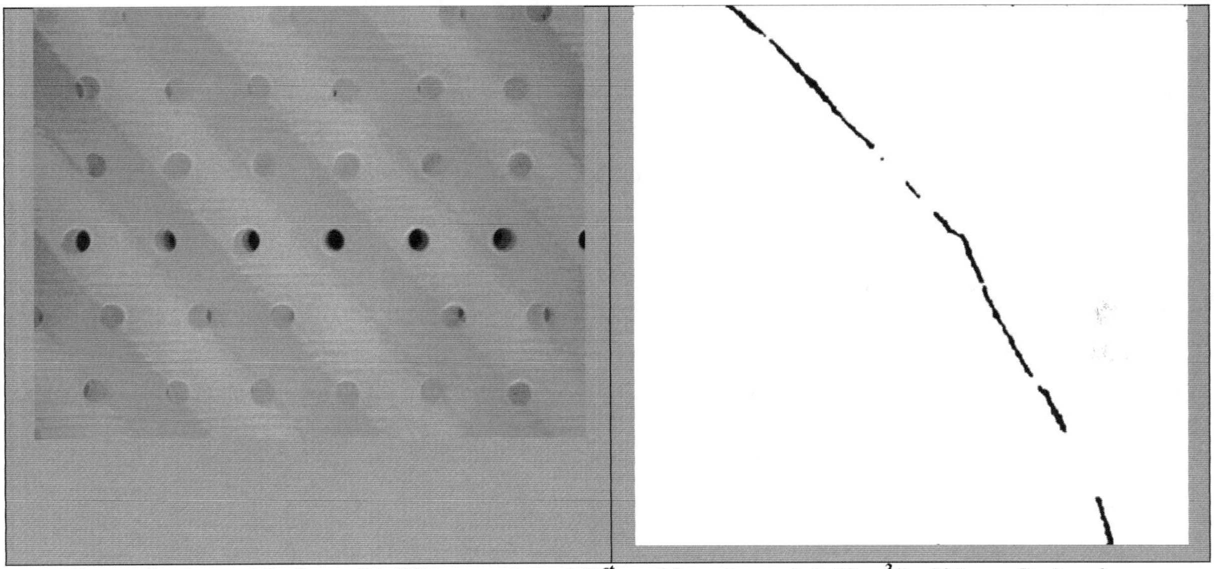

Figure 2: A particle from the fifth experiment (1st packing element, 5.68 m^3/hr [25 gpm]) showing a predominately linear path.

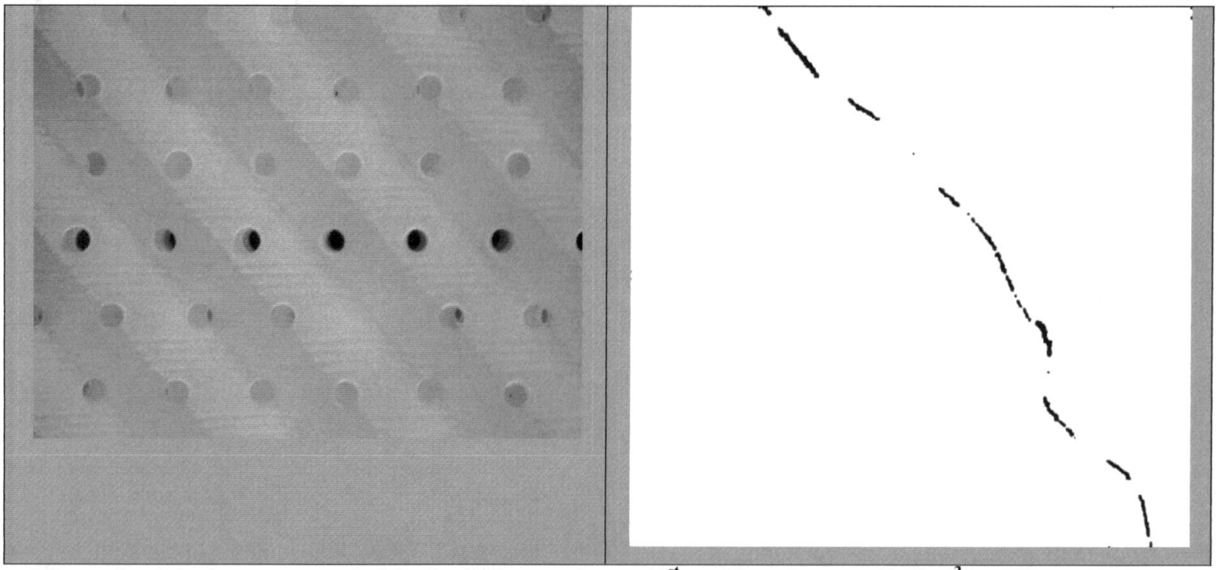

Figure 3: A trajectory recorded in the fifth experiment (1st packing element, 5.68 m^3/hr [25 gpm]) shows minor flow disruptions at the packing interface.

Figure 4 shows a much greater flow disturbance upon entering the packing. The magnitude of the lateral displacement suggests the particle leaves the original channel and flows horizontally through a channel in the adjacent packing sheet (Figure 5). This is followed by a sharp change in direction as the particle seems to enter and flow with a channel in the original packing sheet. Both the horizontal travel and the discontinuous change in velocity would result in high energy dissipation and pressure drop.

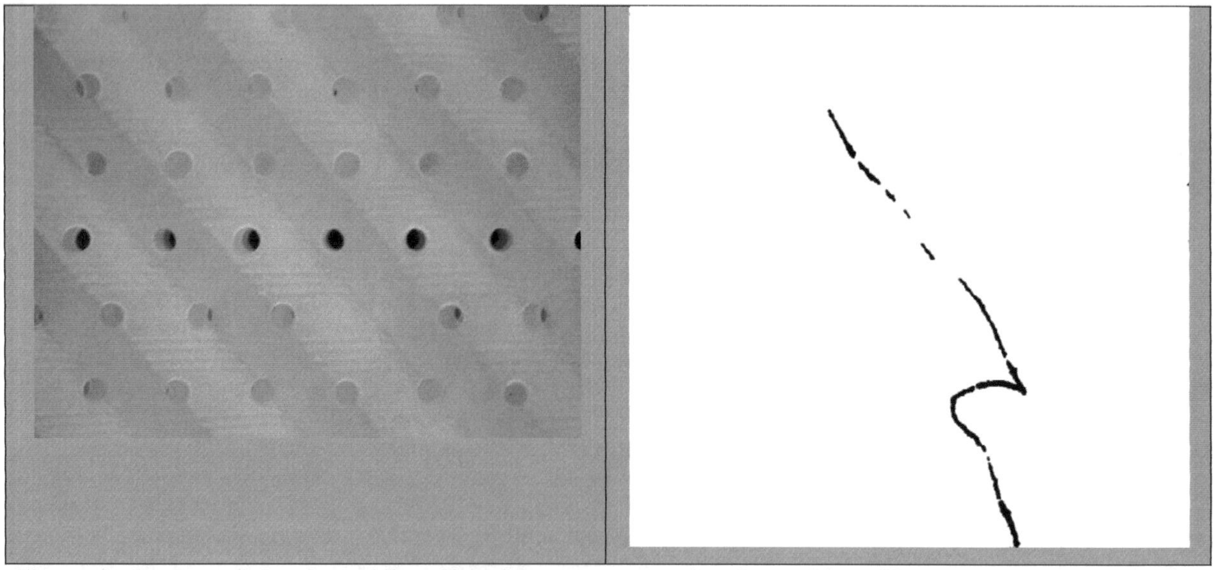

Figure 4: Some particles showed flow disturbance upon entering the packing. Flow through lowest packing element at 5.68 m^3/hr [25 gpm].

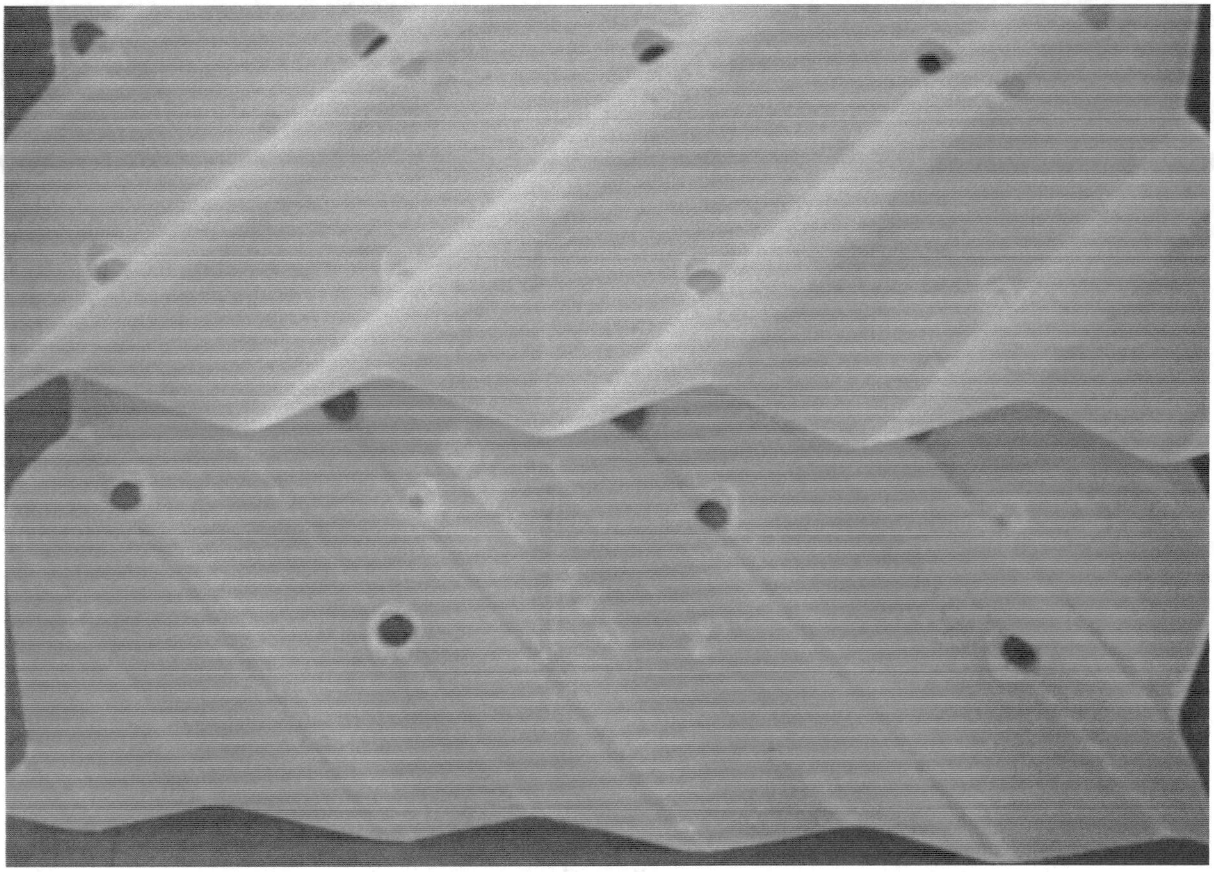

Figure 5: Channels in adjacent packing sheets intersect at 90 degree angles.

Figure 6 depicts a particle path recorded during the seventh experiment. In this path, the particle appears to make a quick transition between channels in adjacent packing sheets. This region is likely characterized by high shear and turbulence. In spite of this, the flow does not exhibit large recirculation. This, coupled with the abrupt horizontal to vertical transition in Figure 4, suggests transitions between adjacent channels do not exhibit wide scale flow disruption.

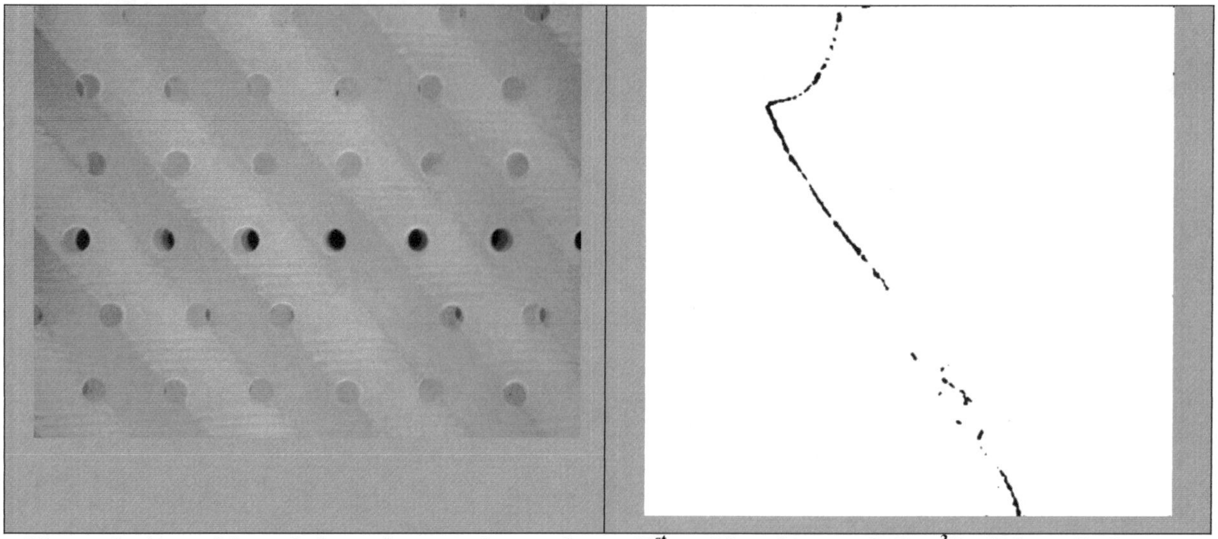

Figure 6: A particle path from the seventh experiment (1st packing element, 5.68 m^3/hr [25 gpm]) shows a particle transition between adjacent channels in the packing.

Figure 7 shows a path which suggests there is a large scale recirculation zone in the middle of the packing. However, the first half of the path shows predominately vertical travel. This likely shows a particle flowing along the wall before encountering an obstruction, such as the wiper band. This obstruction seems to create a separation point which results in a recirculation zone. As the particle exits this zone, it enters the packing and flows smoothly through it.

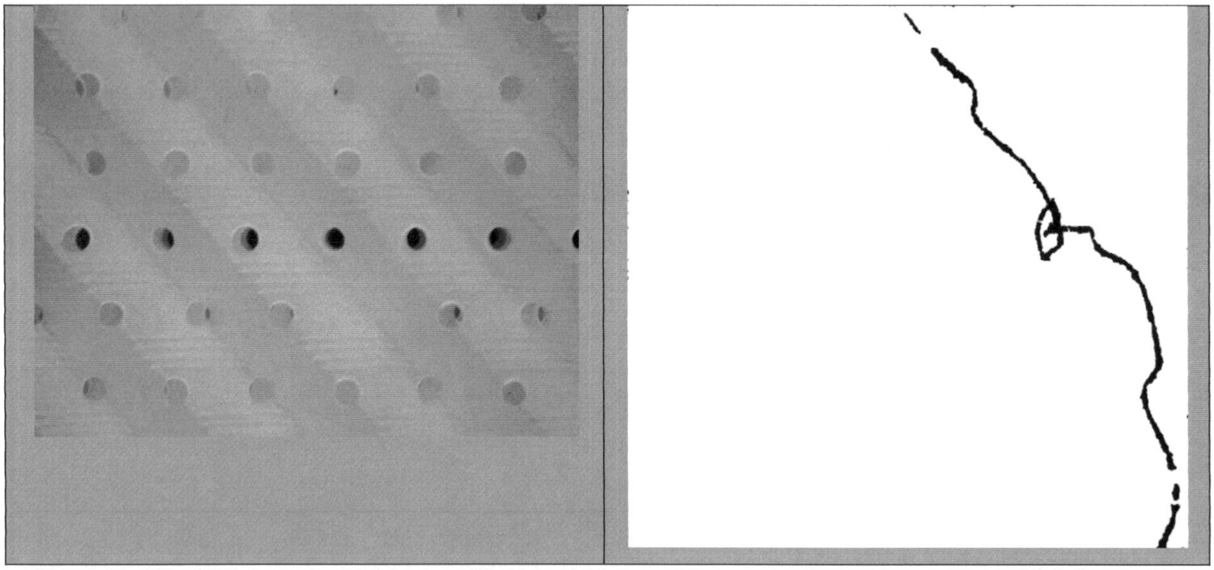

Figure 7: The initial vertical trajectory likely indicates wall flow prior to an obstruction which forces the particle into the packing. Flow at 5.68 m^3/hr [25 gpm] through 1st packing element.

Figure 8 shows the path of a particle as it transitions from the lowest to middle packing elements under a 5.68 m^3/hr [25 gpm] flow rate. Interestingly, this path shows two transitions between packing channels in the lowest packing element with the first transition exhibiting slight negative flow and the second seemingly smooth. The particle exhibits uniform and smooth flow as it leaves the first packing element and enters the second.

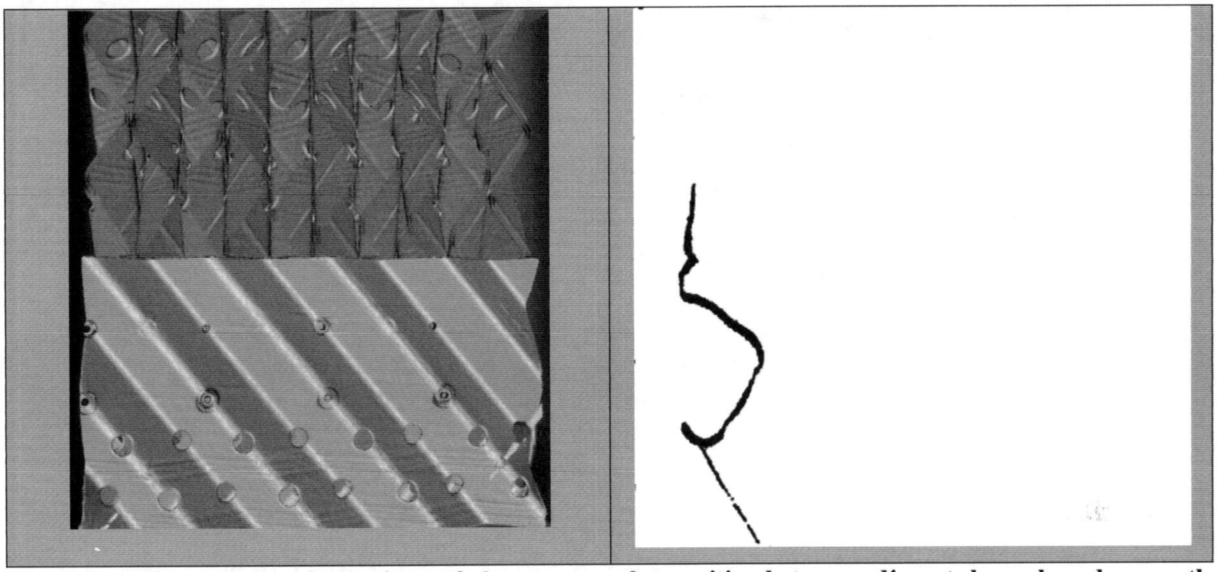

Figure 8: This composite of experiment 8 shows a smooth transition between adjacent channels and a smooth transition through the packing joint. Flow through the packing joint at 5.68 m^3/hr [25 gpm].

Performance at the joint suffered from poor flow integration as well as recirculation and stagnant zones. Figure 9 shows a particle which is shunted towards the wall by the channels in the lower packing. The particle leaves the lower packing element at the joint, but does not enter the next packing element. Instead, the particle flows along the wall and entirely bypasses the packing element. This behavior would result in reduced contacting and limited mass transfer in industrial applications.

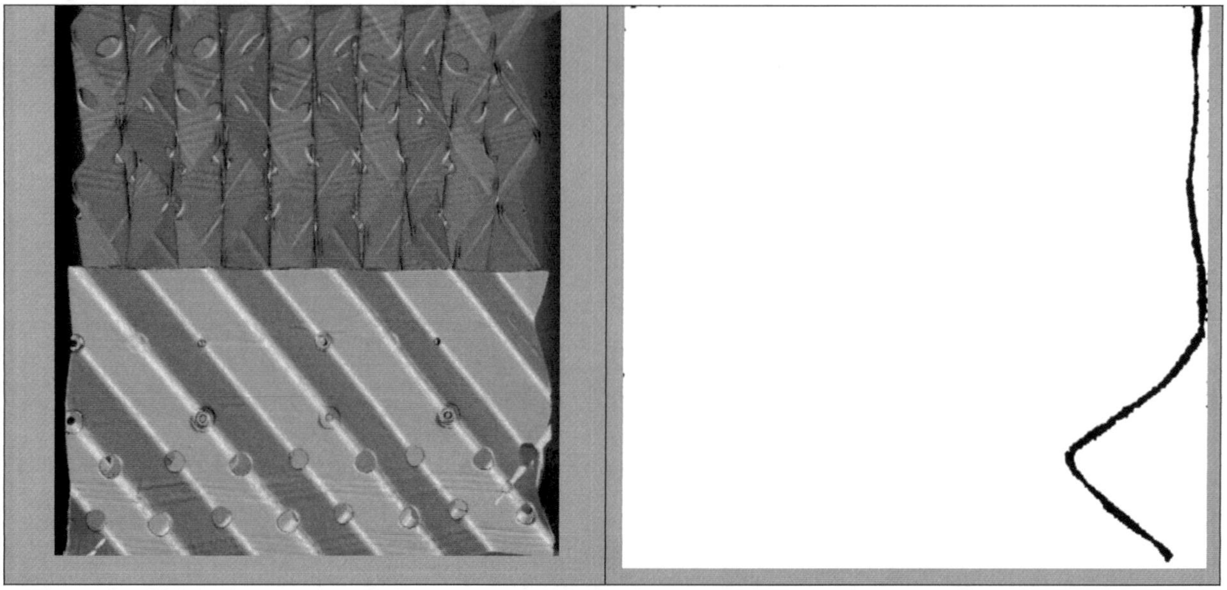

Figure 9: Video of experment 8 shows a particle leave the lower packing element and entering a region of wall flow. Flow through the packing joint at 5.68 m^3/hr [25 gpm].

Negative flow was a common observation at the joint (Figure 10). The particle would enter the second packing element and then reverse direction. Most often the particle would begin flowing up the column again after some horizontal travel. This behavior is suggestive of a non-uniform pressure profile at the entrance to the second packing element and, presumably, all subsequent packing elements. This phenomenon could result in channeling and poor distribution.

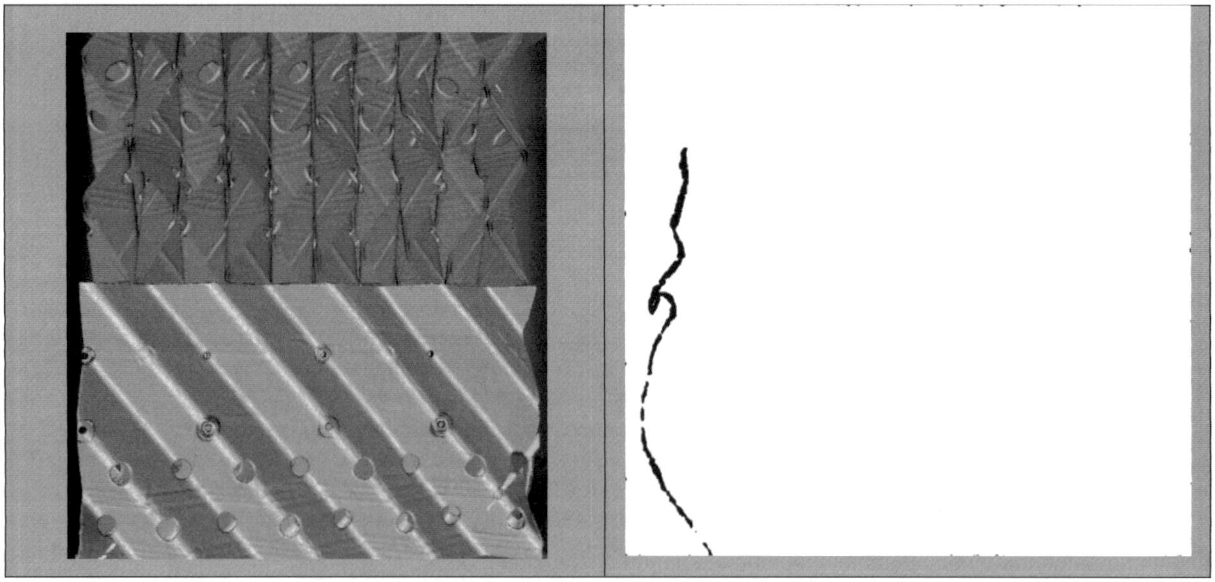

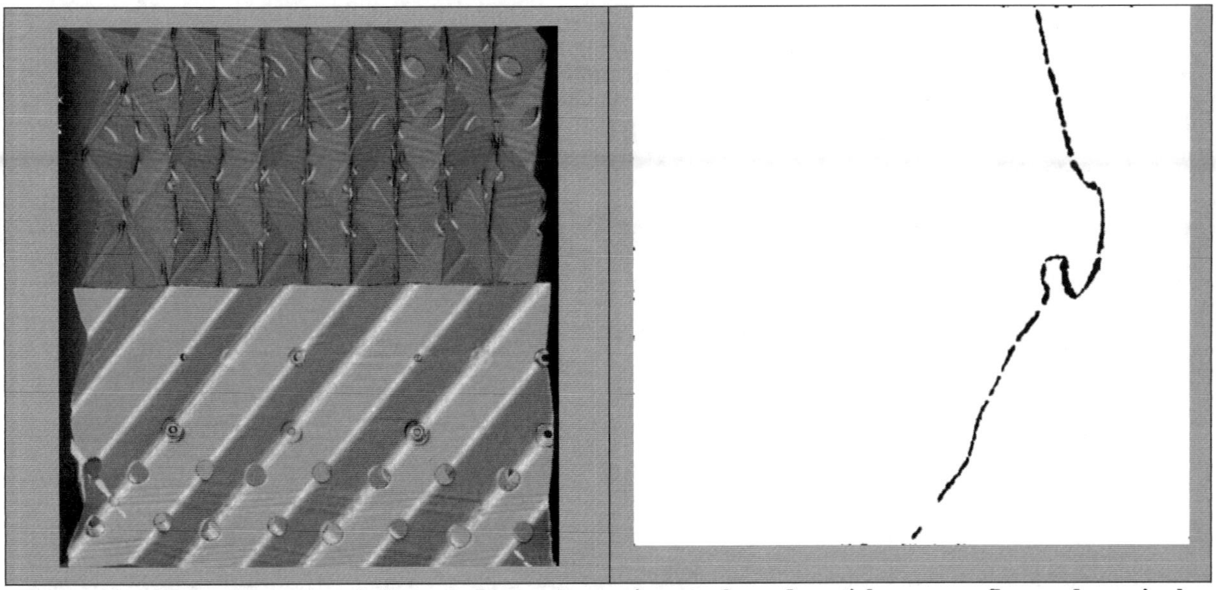

Figure 10: The tenth (top) and eleventh (bottom) experiments showed particles reverse flow and seemingly pass under a packing sheet. Shown is flow through the lowest packing joint at 7.95 m^3/hr [35 gpm].

Figure 11 shows a transition between packing elements. Most notable is the large scale recirculation and corresponding negative flow. This is indicative of separation points forming at the end of the packing sheets. The recirculation would be a significant source of pressure drop and poor contacting performance. Also worth noting is the smooth and uniform nature of the path once it is past the joint which further supports the belief that little flow disruption occurs within the packing elements.

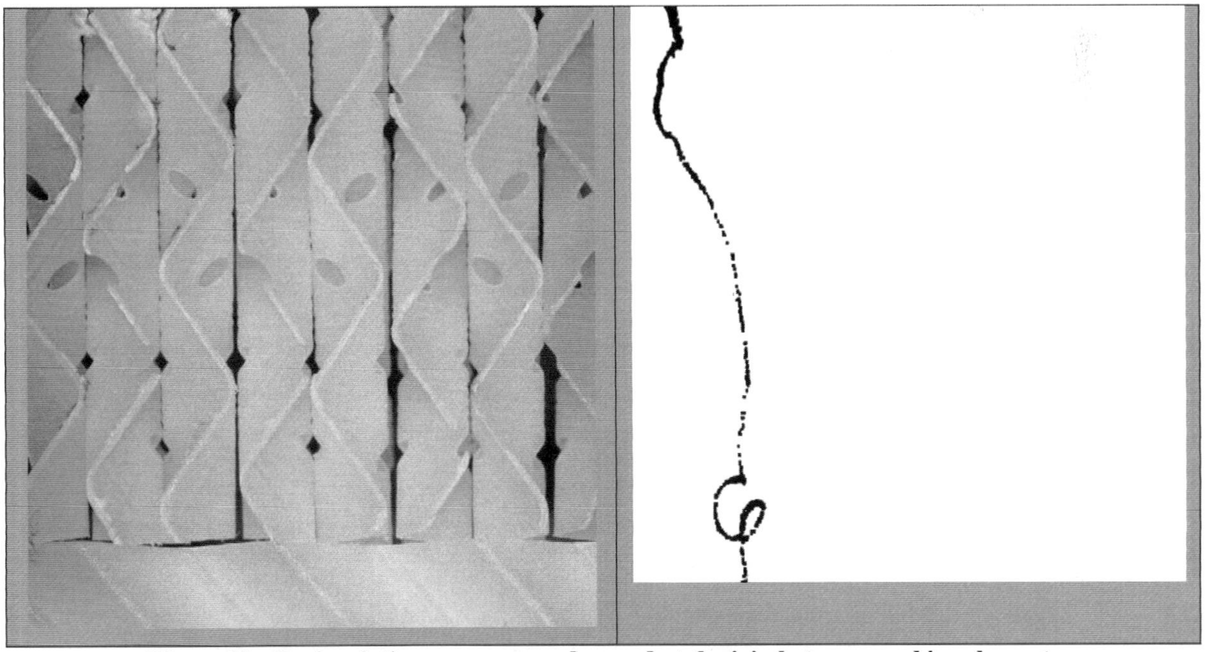

Figure 11: Recirculation zones were observed at the join between packing elements.

3.0 Conclusions

The experimental method was successful in capturing both qualitative and quantitative flow field information inside contacting equipment without optical access. This information is crucial to both focused development of new packing designs as well as validation of existing computational models. The method described above does not suffer from the need to modify equipment for optical access and the size of the equipment is only limited by the power of the X-ray source and size of the detector. This allows the method to be applied more quickly and to a wider range of processing equipment than previous PTV experiments.

The qualitative information captured in this experiment shows that flow within packing elements is largely undisturbed and characterized by linear paths in a single channel. Transitions between adjacent channels were abrupt but smooth and produced little flow disruption. By contrast, flow at the joints was characterized by macro scale recirculation and poor flow integration (wall flow). Improvements in these areas will yield improved contacting and decreased energy intensity. Identifying these features allows packing manufacturers to focus improvements in these areas which will advance packing designs at an accelerated rate.

The quantitative information was shown to accurately reflect the average interstitial velocities. This suggests that the local flow information captured is representative of the velocity variation throughout the packing. This data can be immediately compared to vapor phase flow physics predicted by existing CFD models. Validation of existing models is required to computer aided design of packing to replace extensive and expensive prototyping. The adoption of a CAD method would allow for a reduced research costs and a shorter time to market as well as enhanced performance.

64d Pressure Drop Measurements on Distillation Columns

Tony Cai, A. Shariat, and Mike Resetarits. Fractionation Research, Inc., P.O. Box 2108, Stillwater, OK 74076

Pressure drops are of major importance on distillation/absorption columns. This paper mainly discusses how to correctly measure, interpret and use pressure drop data. The possible causes of incorrect pressure drop measurements are studied including the effects of pressure tap dimensions, locations, and vapor condensation etc. The effect of the static head of vapor on the pressure drop data and column pressures is evaluated. Variations of sectional pressure drops along the column are investigated based on the experimental data obtained from commercial size distillation columns at Fractionation Research, Inc. (FRI). For a packed column, it is found that the spacing between the liquid distributor and the top of the bed affects the overall pressure drop measurements, which is confirmed by a fundamental fluid dynamics analysis.

Key Words: Distillation, tray, packing, pressure drop,hydraulics

Web Page: www.fri.org/

64e INTALOX® ULTRA™ - An Improved Random Packing

Izak Nieuwoudt, Koch-Glitsch, 4411 E 37th Street N, Wichita, KS 67220 and Pat Quotson, Koch-Glitsch, LP, 4111 E 37th Street N, Wichita, KS 67220.

INTALOX® ULTRA™ random packing is a new development that provides improved capacity and efficiency compared to other random packing. Data over a wide range of operating conditions will be presented for INTALOX ULTRA random packing and for other random packing that are commonly used in industry. Case studies will be used to demonstrate the benefits of this new random packing.

64f Raschig Super-Ring - Fourth Generation Random Packing Keeps Customer Exited in Performance

Michael Schultes, Raschig Ring Department, RASCHIG GMBH, Mundenheimerstr. 100, Ludwigshafen, Germany and Simon Chambers, JAEGER PRODUCTS INC., 1600 Airport Freeway, Bedford, TX 76022.

Raschig Super-Ring is now available on the market since app. 10 years and it's time to summaries a success history. Nobody wanted to be first but since Raschig Super-Ring had shown its superior performance in mass transfer efficiency, capacity and pressure drop this random packing becomes more and more used in world wide industries.

The presentation will provide information about grass root designs as well as revamp cases using Raschig Super-Ring. The applications will cover distillation and absorption processes as well as heat transfer zones in fouling, foaming and regular systems.

Beside industrial applications new FRI research results will be shown as well. These research results will outline future applications in driving columns even harder.

64g Design Guidelines for Chemical Treatments in Distillation Columns

Karl Kolmetz[1], Marilyn Blaschke[2], Wai Kiong Ng[3], and Jeff N. Gray[1]. (1) KLM Technology Group, Singapore, Singapore, (2) Baker Petrolite, a Division of Baker Hughes, 12645 West Airport Blvd., Sugar Land, TX 77478, (3) Crystallisation and Particle Science, Institute of Chemical and Engineering Sciences, 1 Pesek Road, Singapore, 627833, Singapore

Distillation is the application and removal of heat to separate hydrocarbons by their relative volatility or boiling points. This necessary addition of heat normally in the feed stream or at the tower bottoms via a reboiler can also lead to other unwanted consequences such as polymerization, corrosion and reverse solubility. The removal of heat can lead to sedimentation, solubility effects, corrosion and precipitation. The concentration of certain constituents by the distillation process can cause corrosion and flow phenomena effects.

A properly designed distillation column can reduce the effects of these consequences, but in certain applications the polymerization, corrosion and other effects are very prominent leading to reduced separation efficiency in the column. This reduced separation efficiency increases the need for column maintenance and unit down time. In these applications a review of tower internal design and process chemical treatments should be initiated. Previously a review of tower internal design for fouling services was published, whereas this article will discuss the proper application and best practices of chemical treatments in distillation columns.

84a Operating Parameters and Selectivity in Batch Reactive Distillation

Wei Qi, Department of Chemical Engineering, University of Massachusetts Amherst, Amherst, MA 01003 and Michael F. Malone, College of Engineering, University of Massachusetts, Dean of Engineering, 125 Marston Hall, Amherst, MA 01003-9303.

Batch reactive distillation (BRD), combines the flexibility of batch process with advantages of reactive distillation, and is an important process development tool for reactive distillation.

This paper is focused on selectivity improvements for improved raw material consumption and lower waste production, emphasizing the impact of operating parameters on selectivity in BRD. We describe the impact in terms of the appropriate dimensionless groups: the reflux or reboil ratio, along with a certain type of Damköhler number (*Da*), which gives the ratio of the process time to the reaction time.

We show that selectivity improvements in BRD can be limited for high values of *Da* and that selectivity is enhanced at the expense of conversion as *Da* is decreased. This conversion loss can be mitigated by increasing the reflux ratio. Consequently, for systems operated at low *Da*, there is an optimum value of reflux ratio that provides the maximum yield.

We show sample results for a hypothetical case of serial isomerization reactions and also a two-reactant system (synthesis of ethylene glycol), in which distillation may have a negative impact on selectivity by causing separation of the reactants. For the latter example, the reboil ratio in a batch stripper has a crucial impact on selectivity. In fact, too low a reboil ratio will cause even lower selectivity than found in a conventional batch reactor. Understanding the impact of these parameters allow us to explore the tradeoffs involved in choosing among alternative operating strategies.

84b Batch Distillation Back-Mixing: Impact on Batch Yields

Steven D. Kohler[1], Laxminarasimhan Padmanabhan[2], Ellen C. Murphy[1], Louis P. Silveri[1], and Manuel R. Florez[1]. (1) Process Engineering, LyondellBasell, Flavors & Fragrances, PO Box 389, Jacksonville, FL 32201-0389, (2) Aspen Technology, 1010 N Virginia Ave NE Apt 6, Atlanta, GA 30306

Key Words: Batch, Distillation, Back-Mixing, Process Performance, Batch Strategies, Industrial Environment, Batch Automation, Process Modeling

Abstract

Batch distillation optimization is a continuous effort in every batch manufacturing facility. Process improvements are driven by increasing operating costs. It is therefore imperative to reduce, and when possible, eliminate wasteful practices in both equipment design and operating procedures.

This paper discusses the unfavorable impact of distillate back-mixing on batch distillation yields. Various common equipment design flaws are shown and design alternatives are presented for handling distillate with minimum back-mixing effect. In addition, a real-life example is presented showing how batch yields were improved by 3-6%.

84c Dynamic Modeling and Optimization of Thermally Coupled Dividing Wall Column Batch Distillation Processes

Girish S. Joglekar, Batch Process Technologies, Inc., 1291 Cumberland Avenue, West Lafayette, IN 47906, Gintaras V. Reklaitis, Purdue University, 480 Stadium Mall Drive, West Lafayette, IN 47907, and Rakesh Agrawal, School of Chemical Engineering, Purdue University, 480 Stadium Mall Drive, West Lafayette, IN 47907.

A novel, dividing wall column batch distillation process has been developed to distill multicomponent mixtures into multiple pure product streams. Unlike in continuous columns, the dividing wall in the proposed batch column extends upto the bottom of the column, creating two chambers in the lower section of the column. Each lower chamber has its own reboiler and heat input. Like a conventional batch column, there is a condenser at the top. An important feature of this process is that it can operate as a conventional batch distillation column if necessary. The dividing wall batch distillation processes are expected to significantly reduce the capital cost as it requires only one column shell. The potential applications are in pharmaceutical, specialty chemical and agricultural industrial facilities that utilize a wide range of solvents.

In this study, a detailed process dynamics model of the thermally coupled columns was constructed. Distefano's quasi steady-state formulation(2) was used for mass balance on all plates, while the rest of the variables were updated after the specified elapsed time. The important variables that affect the column performance are the reflux ratio, cut durations and fraction of liquid stream within the column that is split between the two chambers created by the dividing wall. A mixture of Benzene, Toluene and Ortho-Xylene with mass fractions 0.33, 0.33 and 0.34, respectively, was charged to the column.

The following key operating variables associated with each cut affect the overall column performance: reflux ratio, duration and fraction of liquid stream within the main column that is diverted to the side column. To find the optimum values of the operating variables, a two-level approach was used. The Genetic Algorithm function in Matlab was used to create the members in the population. The fitness function value for each member was computed by simulating the column dynamics. The objective function value of each member in turn was used by the GA to create new population.

For a three component mixture of Benzene, Toluene and Ortho-Xylene, the optimum cycle time for the dividing wall column process is reduced by 34% compared to the optimum cycle time for a conventional column; also slop cut is eliminated leading to environmentally friendly solution.

1. AIChE presentation, 2007 Annual Meeting, paper 211a.

2. AIChE Journal, 14(1), 190-199 (1968).

84d On Finite Speed of Diffusion Effects on Murphree Plate Efficiency

Sharma, Renganathan Kal, Department of Chemical Engineering, Prairie View A & M University, PO Box 519, MS 2505, Room 201A, C. L. Wilson Building, Prairie View, TX 77446

The McCabe and Thiele method for design of multi-stage binary distillation tray tower includes the following main steps:

Step I: Derivation of Operating Line from Mass Balance

Step II: Representation of Equilibrium Line from

Thermodynamic Models and Experiments

Step III: Mass Transfer Effects lumped into a parameter

called Efficiency

Step IV: Step of the number of stages needed to meet

the desired objectives

The real line can be calculated using the lumped parameter called efficiency. The diffusion from the liquid phase to the interface and from the interface to the bulk of the vapor phase is transient in nature. It has been shown (Sharma,K. R., Damped Wave Transport and Relaxation, (2005), Elsevier, Amsterdam, Netherlands) Fick's law of diffusion can be modified to include a ballistic term without violation of second law of thermodynamics. This way the finite speed of mass propagation can be accounted for. In this study, the hyperbolic PDE are solved for in the liquid and vapor phases and the effect of finite speed of mass propagation on the real line is studied. For systems with large relaxation times the concentration can undergo subcritical damped oscillations in concentration.

Improve Column Performance; Operate Closer to the Hydraulic Limit without Flooding

Limited refining capacity strains petroleum processing capabilities on a global scale. Worldwide refining of petroleum fuels cannot keep pace with the growing demand for gasoline and distillate. If one couples refining limitations with today's gasoline and distillate margins one can appreciate the strong financial incentives for refiners to push their plants, specifically distillation columns, to maximize throughput. However, pushing a distillation column to its hydraulic limit, while profitable, may have undesirable consequences.

Flooding in distillation columns occurs when the liquid/vapor traffic inside the tower is sufficient to inhibit the column from making the desired separation. The two primary types of flooding in tray columns are *jet flooding* and *downcomer flooding*, each with its own tendencies. Jet flooding occurs when the liquid in the tower is heavily entrained by the vapor. A downcomer flood occurs when the liquid flow through the downcomer is choked off as result of excessive pressure drop associated with vapor flow and/or liquid flow through the downcomer. If left unchecked liquid holdup propagates up the tower resulting in a loss of separation efficiency, a reduction in bottoms flow, and an increase in the pressure drop. Depending on the severity of the flood, consequences such as throughput reductions, off-spec products, equipment damage, and tower shutdown are common. Therefore, plant operations personnel are reluctant to increase distillation column beyond well-established column delta-pressure limits. Delta pressure is the measurement of the differential pressure between two points in the tower.

Conventional distillation towers generally utilize the pressure drop measurement across the column's internals to approximate the proximity to the hydraulic limit (the flood point). But a column's incipient flood point cannot be directly identified with conventional instrumentation. Once achieved, and if no recovery action is taken, the incipient flood will eventually propagate to a full-blown flood event. In order to avoid pushing the column beyond the [undetectable] incipient flood point, plant operations personnel restrict the severity of column operation by controlling the tower at or below a well established upper delta-pressure limit. This limit represents a conservative operation point at which, or below, the column never floods. Only in this manner can the plant confidently avoid the costly, negative consequences of flooding. Most often, a column's upper delta-pressure limit is established empirically by plant operators and engineers.

Relying solely on delta-pressure as a substitute for a column's hydraulic limit will result in the column being operated sufficiently far away from its true hydraulic limit to avoid a flooding condition. Therefore, the degree to which the tower is optimized, or throughput is maximized, will be constrained by the column's upper delta-pressure limit.

Flooding Predictor™

In contrast to controlling a column to a delta-pressure limit, the Flooding Predictor™ allows a column to be controlled closer to its true hydraulic limit. This is possible because the pattern recognition techniques employed by the Flooding Predictor™ are capable of identifying the onset of flooding in a distillation column.

The Flooding Predictor™ operates on the principle that between five to sixty minutes in advance of a flooding event,- certain column variables experience an oscillation, a pre-flood pattern. This pattern is a very brief, highly repeatable, simultaneous movement among the derivative values of certain column variables. The pattern recognition system of the Flooding Predictor™ utilizes the mathematical first derivative of certain column variables to identify the column's pre-flood pattern(s). While all column variables experience negligible random noise generated from the natural frequency of the process, a subtle pre-flood pattern develops among a sub-set of column variables when the process enters a transition phase between steady-state and flood. The sub-set of column variables is identified empirically through analysis techniques developed to facilitate the identification of a column's unique pre-flood pattern(s). Prior to the flood event, the identified sub-set of derivative values undergo a temporary, simultaneous change in magnitude identifying it as a pre-flood pattern.

These pre-flood patterns were demonstrated to be predictive of flood with a 98+% accuracy in the closed-loop alpha versions commissioned in a large mid-west refinery. The identification of pre-flood conditions and the development of a mathematical model to represent the pre-flood pattern led to the solution of serious column flooding problems.

Directionally, the pre-flood changes among some column variables are well known. For example, as a column approaches incipient flood, liquid inventories on the trays increase momentarily, resulting in a temporary decrease or erratic behavior in bottom level. Furthermore, liquid hold-up will produce a temporary increase in delta-pressure, as measured in the vicinity of the increased liquid volume.

However well understood these process changes appear to be, they go undetected. All pre-flood excursions are well within the normal operating limits of the column. Thus there is no indication in the normal view of the process that a problem might be developing. For example, the bottom temperature does not increase so that a high temperature alarm is triggered. The same is true of the column delta-pressure, bottom and overhead flow. During a pre-flood event none of the variables move sufficiently far away from steady state to trigger either a high or low alarm point, or provoke the interest of the console operator. The same cannot be said of derivative values. The pre-flood change in magnitude of derivative values is identifiable and easily alarmed to warn of an impending problem.

Research at the Separations Research Program, The University of Texas at Austin

The results of the Flooding Predictor™ test performed at the University of Texas were very encouraging. The technology was able to prevent flooding in a trayed distillation column with a high degree of success. It was found that capacity may be increased by as much as 6%. This advantage will translate into energy and capital savings by preventing off-spec products, improving capacity (for certain applications) without the need for column internal replacement, and giving plant personnel confidence to operate the column near its flood point with a much smaller chance of actually going into a flood.

The control strategy for the distillation column maintained distillate and bottoms product composition by controlling the temperature at the top and bottom of the tray section, with reflux flow and steam flow to the reboiler, respectively. To determine the amount of improvement generated by the Flooding Predictor™, the capacity of the trays and packing first had to be measured without the Flooding Predictor™ enabled. Next, the critical constants (values which triggered the Flooding Predictor™ to reduce column loading) for the derivative variables had to be determined. Finally the Flooding Predictor™ had to be enabled and the critical constants fine-tuned through empirical testing.

Determination of the baseline flood point was achieved by placing the column into composition control mode and slowly increasing the feed until the column flooded. This process was repeated several times to ensure the flood point had been accurately ascertained. Derivative values were recorded and these data served as the baseline critical constants.

After determining where the flood point occurred, the column was returned to an unloaded state and the Flooding Predictor™ was enabled. Feed flow to the column was slowly increased in 2% increments and the column entered the loading region where instabilities were more likely to push the column to flood. The Flooding Predictor™ algorithm compares user-entered critical constants to the actual process variable derivatives and relaxes the severity of column operations via a 5% reduction in reboiler duty when all Flooding Predictor™ criteria are met simultaneously.

Baseline tests showed the maximum sustainable feed rate to the column was 7.03 gal/min. Increasing the feed rate by 2% to 7.11 gal/min resulted in a column flood. Enabling the Flooding Predictor™ allowed the column to be operated in a stable fashion at 7.45 gal/min. This equates to a 6.22% improvement in sustainable feed rate.

Condensed from a paper originally coauthored by:
George E. Dzyacky *2ndpoint, LLC*
A. Frank Seibert *University of Texas at Austin*
J. Christopher Lewis *University of Texas at Austin*

84f Practical Control of Dividing-Wall Columns

Jens P. Strandberg, Chemical Engineering, Norwegian University of Science & Technology, Trondheim, Norway and Sigurd Skogestad, Chemical Engineering, Norwegian Univeristy of Science and Technology, Sem Sælandsvei 4, Trondheim, 7491, Norway.

In this paper, we have studied the practical implementation of stabilizing control for dividing-wall distillation columns. The focus is on how to operate the column in practice, where the goal is to achieve acceptable operation using simple control policies. In particular, we investigate the importance of properly adjusting the liquid split ratio, which determines the relative amount of reflux to the two sides of the dividing-wall The examples include the 4-product Kaibel column and the more well-known 3-product Petlyuk column. In the study, we assume that the objective is to maximize the purity of all product streams, and we show that setting the correct liquid split ratio is essential in achieving the potential purities. Control configurations with varying number of temperature loops have been tested and compared. We show that the liquid split can be used to control a temperature in the prefractionator section and thereby reduce the sensitivity to disturbances. Adjusting the liquid split is particularly important in reducing the column's sensitivity to the vapor split ratio. We also show that for high-purity columns the need to adjust the liquid split online is even greater.

84g Optimizing Energy- and Composition Control of Dividing Wall Columns

Ivar J. Halvorsen[1], Maryam Ghadrdan[2], Mohammad Shamsuzzoha[2], and Sigurd Skogestad[2]. (1) Ict, SINTEF, O.S Bragstads plass 2D, Trondheim, 7465, Norway, (2) Chemical Engineering, Norwegian Univeristy of Science and Technology, Sem Sælandsvei 4, Trondheim, 7491, Norway

Distillation is the most widely used separation method in the chemical process industry and is a large energy consumer among process units. This fact has provided a continuous incentive towards the search for more energy-efficiency distillation systems. The divided-wall distillation column is a fully thermally coupled column, capable of separating three products in a single column shell with a single reboiler. Many papers discuss the steady-state design issues and propose heuristic and rigorous design optimization methods, but the dynamic control of the divided-wall column has been explored in few works. This paper addresses the task of simultaneous optimization of the energy consumption and control of the product composition of a divided wall column (DWC). Only a few papers have addressed both control and optimization at the same time. In a series of papers by Halvorsen and Skogestad (1999,, 2003, 2004) the steady-state relationships for fully thermally coupled arrangements or Petlyuk arrangements (where DWC is a subclass) has been analysed in detail. These results can be used to design the basic control structure for a given case. Then ordinary controllability analysis can be applied to find a practical solution for the total control and optimization task.

Here we suggest a basic general control structure based on 5-point control. This solution will be feasible for all types of ternary mixtures, provided that a sufficient number of separation stages have been designed for the given specifications and feed. In a majority of practical cases this general control structure can be simplified. How, and when depends on the feed properties and the purity requirements and value of the products. Thus, even one or two-point control may be sufficient for some applications. The key to the solution is to know how and when it is required to do on-line adjustment of the liquid split at top of the dividing wall or the vapour split at the lower end, or even both, in order to maintain minimum energy operation. The control loops for product composition must also be in operation. Simulations of the column were performed. The effect of reflux rate, number of column stages and the effect of the vapour split were investigated. The impact of some variables such as the number of trays in various sections and non-optimum design values of the vapour split on controllability and rangeability of the divided-wall column were also studied.

References:

E.A. Wolff and S. Skogestad, "Operation of integrated three-product (Petlyuk) distillation columns", Ind. Eng.Chem.Res., 34, 6, 2094-2103 (1995).

I.J. Halvorsen and S. Skogestad, "Optimal operation of Petlyuk distillation: steady-state behavior", J. Proc. Control, 9, 407-424 (1999).

I.J. Halvorsen and S. Skogestad, "Minimum Energy Consumption in Multicomponent Distillation. Part 1. Vmin Diagram for a Two-Product Column", Ind. Eng. Chem. Res., 42 (3), 594-604 (2003). "Part 2. Three-Product Petlyuk Arrangements", Ind. Eng. Chem. Res., 42 (3), 610-615 (2003). "Part 3. More Than Three Products and Generalized Petlyuk Arrangements", Ind. Eng. Chem. Res., 42 (3), 616-629 (2003).

I.J. Halvorsen and S. Skogestad, "Shortcut analysis of optimal operation of Petlyuk distillation", Ind. Eng. Chem. Res., 43 (14), 3994-3999 (2004).

97a Design Guidelines for Extractive Distillation Columns

Karl Kolmetz[1], Phaik Sun Cheah[2], Wai Kiong Ng[3], Jeff N. Gray[1], and Tim Zygula[1]. (1) KLM Technology Group, Singapore, Singapore, (2) Titan Petrochemicals, Pasir Gudang, Malaysia, (3) Crystallisation and Particle Science, Institute of Chemical and Engineering Sciences, 1 Pesek Road, Singapore, 627833, Singapore

Extractive Distillation is an important tool for the separation of isomers and close boiling species. An extractive distillation solvent is added to the column increasing the relative volatility of the close boiling species allowing distillation to be utilized. Several applications of extractive distillation have been successfully commissioned. Two of the applications are aromatics and butadiene which will be discussed.

Design concerns for extractive distillation include fouling potential, corrosion, multiple liquid phases and foaming. The design of the tower internals must address each of these issues for the application to be a success and have a reliable on steam time between planned outages. Each of these issues will be discussed for the each application.

97b Synthesis and Quick Identification of Low-Energy Sequences for Multicomponent Distillation

Vishesh H. Shah and Rakesh Agrawal. School of Chemical Engineering, Purdue University, West Lafayette, IN 47906

Industrially, distillation is usually the preferred method for separation of multicomponent mixtures. To separate an N-component feed into essentially N pure product streams (N > 3), we can choose from several distillation configurations. These configurations carry out the same overall process; however they differ significantly in total cost. Choosing the optimal distillation sequence is thus an important and challenging problem.

The first step for solving this challenge is to generate a useful search space of distillation configurations. It has been shown that distillation configurations that use more than (N-1) columns for an N-component separation should be omitted from the search space because they are never optimal. We have thus developed a computationally efficient method to generate a search space of exclusively the (N-1) column configurations. We have also extended this method to include configurations with thermal coupling in the search space.

We then focus on minimizing energy consumption since operating cost usually dominates capital cost for large industrial applications. However, since the size of the search space grows exponentially as number of components in the feed increases, we need a quick screening method to identify attractive low-energy sequences because we cannot evaluate each configuration rigorously.

Therefore, the goal of this work is to describe a procedure to identify low-energy sequences by a short-cut screening of the search space. Our formulation enables us to automatically apply Underwood's equations for all the configurations, instead of formulating different problems for each configuration. In this way, we can quickly identify optimal distillation configurations. We can then evaluate the configurations obtained by the quick-screening method more rigorously. The power of our method is reflected in the fact that when applied to mature commercial applications such as crude distillation and ethylene separation, we have been able to identify a number of distillation configurations with potential energy savings in excess of 10%.

97c Optimizing the Design of an Internally Heat Integrated Distillation Column

Mamdouh Gadalla[1], Laureano Jimenez[1], and **Zarko Olujic**[2]. (1) Chemical Engineering Department, University Rovira i Virgili, Av. Països Catalans, 26, Campus Sescelades, Tarragona, 43007, Spain, (2) Process & Energy, Delft University of Technology, Leeghwaterstraat 44, Delft, 2628 CA, Netherlands

Distillation columns that utilize internal heat integration (HIDiCs) were found to offer potential for minimization of energy requirement, which is of particular importance for close boiling separations. Conceptual design and proof-of-operability of these new column configurations have been demonstrated and there are already a considerable number of publications around addressing mainly experimental studies, performance simulation, process dynamics and control aspects. In this work, design insights are provided as well as an effective design tool, incorporating pinch analysis principles, that has been built into a commercial software package to enable a rigorous approach to design and analysis of HIDiCs. This allows detailed insight into performance characteristics and optimization of the basic design of HIDiCs. The usability of this tool is demonstrated; indicating that incorporating a flash drum in the given case can lead to both energy and cost more efficient design of a HIDiC.

97d Efficiency Comparison of Distillation Schemes: Heat Integrated Distillation Column (HIDiC) and Heat Pump Systems

Anirudh A. Shenvi[1], Rakesh Agrawal[1], and D. Michael Herron[2]. (1) School of Chemical Engineering, Purdue University, 480 Stadium Mall Drive, West Lafayette, IN 47907, (2) Air Products and Chemicals, Inc., 7201 Hamilton Boulevard, Allentown, PA 18195-1501

Distillation is the most important separation unit operation in process industries. Conventionally, heat is supplied at a higher temperature in the reboiler and is rejected at a lower temperature in the condenser. If the heat rejected in the condenser is not utilized, then it generally leads to inefficiency. Increased emphasis on energy efficient processes has led to a renewed interest in intensifying this unit operation.

Intensification can be achieved by upgrading the heat rejected in the rectifying section and transferring it to the stripping section. There are various options available to achieve this objective. Heat integrated distillation column (HIDiC) scheme performs this transfer of heat from the rectifying section to the stripping section throughout the length of each section by operating the rectifying section at a higher pressure than the stripping section. Whereas, in a heat pump (direct vapor recompression) scheme, a one point heat transfer operation can be used wherein heat only from the top of the rectifying section is upgraded and is transferred to the bottom of the stripping section. Also, by using a heat pump system incorporated with one intermediate reboiler and/or one intermediate condenser, heat integration at a second point can be achieved. The advantage of these latter arrangements is that one does not need HIDiC hardware to continuously transfer heat from the rectifying section to the stripping section. This simplifies the design of the distillation column and associated equipment.

In the literature, HIDiC system has been compared against conventional distillation columns and been shown to provide considerable energy savings. However, there is a need to compare the HIDiC system against alternative heat pump schemes. Thus, the purpose of this work is to systematically compare the performance of the HIDiC scheme with alternative heat pump systems and recognize its merits/demerits. Initial results show that under reasonable operating conditions, a majority, if not all, power benefits of HIDiC can be achieved by use of simple one or two stage heat pumps between the rectifying and the stripping sections.

97e Design and Analysis of An Externally Heat-Integrated Distillation Column

Huang Kejin[1], Wang Yun[2], Haisheng Chen[2], Guosong Wu[2], Xingxing Zhang[2], and Shaofeng Wang[2]. (1) College of information Science and Technoilogy, Beijing University of Chemical Technology, No. 15, Beisanhuandonglu, Chaoyang District, Beijing, 100029, China, (2) College of Information Science and Technology, Beijing University of Chemical Technology, No. 15, Beisanhuandonglu, Chaoyang District, Beijing, 100029, China

In this presentation, we report our recent research outcomes on an externally heat-integrated distillation column (EHIDiC). Although EHIDiC is in principle quite similar to an internally heat-integrated distillation column (IHIDiC), they are actually very different in process structure, where heat integration is arranged between two different distillation columns in the former and the same distillation column in the latter. EHIDiC can, however, show advantages than IHIDiC in not only energy efficiency but also capital investment, and our research is aimed to exploit these potentials in terms of an ethylene distillation column separating a closing-boiling C2 mixture. The results obtained so far indicate explicitly the general static characteristics of the EHIDiC.

Study on Power consumption of a Novel HIGEE —— Rotating Zigzag Beds

Li yumin, Yu yunliang, Xu zhichao, Ji jianbing, Li xiaohua

College of Chemical Engineering & Materials Science, Zhejiang University of Technology, 310032 Hangzhou, China

Abstract: Using centrifugal force to intensify mass transfer, the HIGEE(high-gravity) units with high performance have been applied widely to chemical processing industries. And the rotating zigzag bed(RZB) is a novel Higee. The rotor of RZB consists of stationary disc and rotating disc, forming zigzag channels for liquid-gas flow and mass transfer. The power consumption of HIGEE unit was important, but seldom studied. The shaft power consumption of RZB was modeled to be the sum of the power requirement for the flow of liquid, for the frictional windage drag of the rotor, for the bearing friction, and for the flow of gas. A theoretical analysis was developed to predict the power requirement for liquid film flow on the rotating disc as two-dimension ideal liquid flow without gas flow using Euler's turbine equation. And the power requirement for real liquid flow with gas flow was equal to the power for ideal liquid flow without gas flow multiplied by the coefficients of K_1 and K_2. The experiments were performed in a RZB unit with the rotor of 600mm-diameter using air-water system. The power consumption with various gas rate, liquid rate and rotational speed was measured. It is shown that the power requirement for the flow of gas was negligible, and the coefficients of K_1 and K_2 depended on the liquid rate, gas rate and rotational speed. Two correlations for the coefficients of K_1 and K_2 were proposed using experimental data, and agreement between the predicted values and the experimental values was good. The power requirement for the rotor frictional windage drag and the bearing friction, only depended on rotational speed, was also correlated. Therefore, an integrated correlation for calculating the shaft power consumption of RZB was obtained.

Key words: Rotating zigzag bed, Rotating packed bed, Power consumption, Liquid film flow, Gas-liquid, Distillation

1. Introduction

As one of the process intensification(PI) technologies, the HIGEE(high-gravity) or the rotating packed bed(RPB) patented by Imperial Chemical Industries(ICI)(Ramshaw, C. 1981, Mallinson. R.H. 1982) has been widely applied to chemical processing industries, even in making nano-material(Chen, J.F. 2000, Chen, J.F. 2000), micromixing of two liquid streams(Yang, H.J. 2006), continuous biosorption(Das, A. 2008), trickle-bed reactors(Dhiman, S.K. 2005, Niqam, K.D.P. 2005), and so on.

This technology uses centrifugal fields as stimulants for process intensification. The advantage of a high centrifugal force, hundreds of times the terrestrial gravity, include the use of packing with a large surface area, formation of thinner liquid film over the packing leading to the enhanced mass transfer coefficient and the height equivalent of a theoretical plate (HETP) as low as 1-2cm, improvement of the gas-liquid slip velocity resulting in the reduced tendency of flooding, and lower static holdup, etc. As a result, the size of the equipment is reduced by 2-3 orders of magnitude as compared with the conventional packed columns, with lower capital and operating cost.(Munjal, S. 1989, Rao, D.P. 2004)

The fundamental studies on the hydraulic characteristics of the rotating packed bed have been carried out in many literatures including the pressure drop(Keyvani, M. 1989, Singh, S.P. 1992, Guo, F. 1997, Kumar, M.P. 1990, Kelleher, T. 1996, Zheng, C. 2000, Lin, C.C. 2002), liquid holdup(Basic, A. 1995, Lin, C.C. 2000, Burns, J.R. 2000), flooding(Lockett, M.J. 1995, Singh, S.P. 1992), the residence time distribution(RTD)(Keyvani, M, 1989, Guo, K. 2000), visual studies of liquid flow(Burns, J.R. 1996, Guo, K. 2000), and so on.

However, power consumption of rotating packed bed was seldom studied. Keyvani(1989) proposed that the total power of rotating bed was classified into four parts: (1) the power required for the acceleration of the liquid to the tip speed of the rotor; (2) the power consumed in overcoming the frictional windage drag of the rotor; (3) the power required for the bearing friction; and (4) the power consumed in friction losses as the liquid passes through the bed. In addition, there was a low power requirement associated with the vapor flow through the rotor bed. Singh(1992) proposed an empirical equation that the power consumption of rotating packed bed is the sum of the power for all frictional losses and the power required to accelerate the liquid entering the packed bed to the rotational speed at the outer radius.

It is very necessary to study the power consumption of rotating packed bed, because the rotating packed bed is rotary equipment with extra power consumption as compared with the conventional packed columns as a static equipment.

2 Rotating zigzag bed

The rotating zigzag bed(RZB) is a new kind of HIGEE(Ji,J.B. 2004, Ji,J.B. 2008, Wang, G.Q. 2008), as shown in Fig.1. The concentric circular baffles fixed to the rotating disc coupled with those fixed to the stationary disc forms the rotor of RZB. The upper stationary disc is fixed to the RZB casing, while the lower rotating disc is fixed to a vertical rotating shaft driven by a motor. The upper portion of the rotating circular baffles is perforated, such as perforations and shutter openings, through which all liquid and part of gas flow. There exit no any openings on the stationary baffles. The gap between the rotating and adjacent stationary baffles is about 10~30mm, whereas the gap between the rotating baffles and the upper disc is about 10~15mm, as well as between the stationary baffles and the lower disc by about 20~30mm. The dynamic seal, denoted by Symbol 13 in Fig.1, is used to prevent the liquid leakage from the casing to the environment.

The gas is tangentially introduced into the RZB casing from the gas inlet, flowing with zigzag radially inward

Correspondence to: Professor Ji Jianbing, College of Chemical Engineering & Materials Science, Zhejiang University of Technology, Hangzhou, 310032, Zhejiang Province, People's Republic of China. E-mail: jjb@zjut.edu.cn through the rotor of the RZB due to the pressure-driving force. As shown in Fig.2, the gas flows upwards through the gap between the stationary and rotating baffles, flowing inwards between the rotating baffles and the upper stationary disc for most of the gas, and inwards through the perforations on the rotating baffles for some of the gas. And then, the gas flows downwards, flowing inwards between the stationary baffles and the lower rotating disc. Finally, the gas emerges at the eye of the casing through the gas outlet.

The liquid enters axially through the liquid inlet at the eye of the casing, is caught up on the rotating disc as the liquid film, and is whirled tangentially and radially outward due to the centrifugal force, until it arrive at the rotating baffle, as shown in Fig.2. And the liquid climbs on the wall of the baffle as the thin film under the influence of the centrifugal force, being dispersed into a shower of fine droplets by perforations on the rotating baffles. The droplets fly outwards with high velocity. Some of them form finer droplets entrained upwards and inwards by the gas, and most of them impact on the wall of the stationary baffle, forming the thin film and falling from the bottom of the stationary baffle. The falling liquid is torn to liquid droplets again due to the action of the gas blow, flying inwards by the gas and falling onto the rotating disc, which is denoted by Symbol 7 in Fig.2. Again, the liquid was whirled tangentially and radially outward until it arrives at the next rotating baffle. Finally, the liquid leaving the rotating disc is collected by the casing wall and flows out of the RZB unit through the liquid outlet at the bottom of the casing.

The gas-liquid intimate contact in the annular space between the rotating and stationary baffles and the mass-transfer occurs (Wang, G.Q. 2008). Hence, there exist two regions. One is the annular region between the inner stationary baffle and the outer rotating baffle as the film-flow region without mass-transfer, the other is the annular region between the inner rotating baffle and the outer stationary baffle as the mass-transfer region. Compared with the conventional RPB, RZB may performance a continuous distillation with intermediate feed in only one unit of RZB, due to the stationary upper disc where the intermadiate feed is accomplished(Wang, G.Q. 2008).

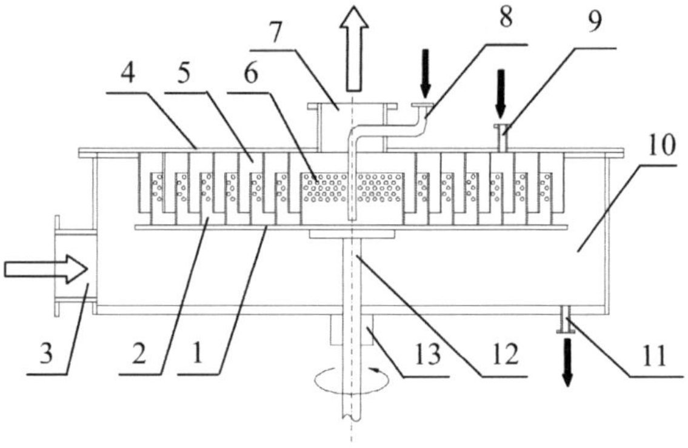

Fig.1. Sketch of rotating zigzag beds(RZB)
1-rotating disc; 2-rotating circular baffles; 3-gas inlet; 4-stationary disc; 5-stationary circular baffles; 6-perforations on the rotating baffle; 7-gas outlet; 8-liquid inlet; 9-intermadiate feed; 10-casing; 11-liquid outlet; 12-rotating shaft; 13-dynamic seal

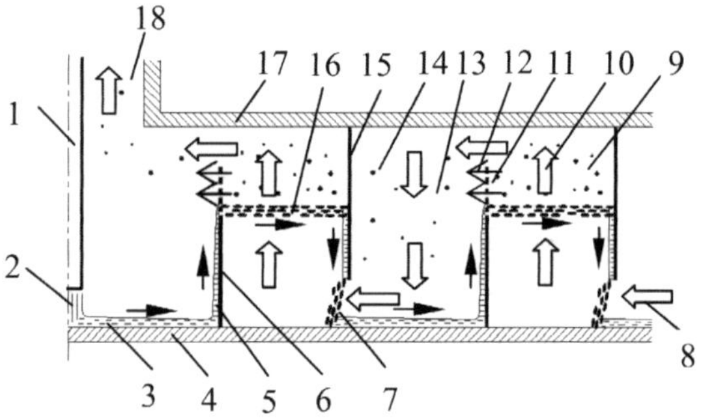

Fig.2. Flow of gas and liquid in rotating zigzag bed(RZB)
1-liquid inlet; 2-liquid; 3-liquid film on the rotating disc; 4-rotating disc; 5-liquid film on the rotating baffles; 6-rotating baffles; 7-liquid droplet flying inwards by gas; 8-gas; 9-mass-trasfer region; 10-zigzag flow of gas; 11-perforations on the rotating baffles; 12-gas through perforations; 13-film-flow region; 14-entrained liquid droplet by gas; 15-stationary baffles; 16-flying liquid droplet; 17-stationary disc; 18-gas outlet

The power consumption of RZB was much greater than that of the conventional RPB, because the kinetic energy of the liquid from the liquid acceleration under the centrifugal force disappear when the liquid as droplets impact on the wall of the stationary baffle with high velocity. Therefore, it is very necessary to study the power consumption of RZB especially. The power required for the acceleration of the liquid is based on the liquid film flow on the rotating disc. Matsumoto(1973) presented an approximate solution for the liquid flow on a rotating disc using polynomial approximations, giving a fair prediction of the liquid film thickness. Munjal(1989) developed the equations for the film thickness and the velocity profiles using the momentum integral method, suggesting that the film height depends on the initial liquid velocity, from which the predicted velocity profiles is applicable for the convective diffusion model. Lin(2000) modified the equations of Munjal, predicting the velocity profiles of the liquid film on the rotating disc for the case of gas flow. Chen(2005) applied the equations of Munjal to a power-law fluid.

3 Model for power consumption of RZB

According to the analysis of Keyvani(1989), the shaft power, N_B, is

$$N_B = N_L + N_W + N_{bf} + N_G \qquad (1)$$

where N_L is the power requirement for the flow of liquid through the rotor including the liquid acceleration and friction losses; and N_W, proportional to $\omega^{1.5}$, is the power consumed in overcoming the frictional windage drag of the rotor; and N_{bf}, proportional to ω, is the power required for the bearing friction, and N_G is the power requirement for the flow of gas through the rotor.

N_L is the power requirement for the liquid flow on which the gas flow acts, and N_{LR} is the power requirement for the liquid flow in the absence of gas flow. Because of the drag of the gas flowing inwards on the surface of liquid film flowing outward and the entrainment of liquid droplets by the gas, N_L will be greater than N_{LR}, which is given by

$$N_L = K_1 N_{LR} \qquad (2)$$

N_{LR} is the power requirement for the real viscous liquid flow without gas flow, and N_{LD} is the power requirement for the ideal liquid flow with the viscosity of zero without gas flow, so it is given by

$$N_{LR} = K_2 N_{LD} \qquad (3)$$

Now a model is developed to predict N_{LD}. It is assumed that

(1) Liquid film flow on the rotating disc is the two-dimensional ideal liquid flow without friction.

(2) The circumferential velocity of the liquid film, u, is equal to that of the rotating disc and baffles.

(3) The velocity of the liquid leaving the bottom of the stationary baffles is zero.

(4) The liquid film spread uniformly over the whole surface of the rotating disc without any dry patch.

The N_{LD} is calculated using the resistance torque, T, of the liquid film on the rotating disc and baffles. And the resistance torque of the liquid film on the rotating baffles is zero, because the circumferential velocity of the liquid film on the rotating baffles is equal to that of the rotating baffles. Hence, the N_{LD} is calculated only according to the liquid film on the rotating disc, because the circumferential velocity of the liquid film varies with radius.

The liquid film flows on the rotating disc is treated in a moving cylindrical polar coordinates system rotating at the angular speed, ω, in the counterclockwise direction, as shown in Fig.3(a) and (b). The coordinate system is defined with the radial distance r from the disc axis, the distance z along the disc axis, and a rotation angle θ about the disc axis. Both centrifugal acceleration, a_{ce}, and Coriolis acceleration, a_{co}, are exerted on the liquid particle moving on the rotating disc. The magnitude of centrifugal acceleration, a_{ce}, is $\omega^2 r$, which direction is along the radial direction; and the magnitude of Coriolis acceleration, a_{co}, is $2\omega \cdot w$, which direction is normal to the tangential direction of the relative streamline or pathline, as shown in Fig3(b).

In the annular film-flow region i on the rotating disc between one stationary and the adjacent outer rotating baffles, the liquid particle makes relative motion with relative velocity, w, at the circumferential motion of the rotating disc with circumferential velocity, u, so the absolute motion with absolute velocity, v, is the resultant of the relative motion and the circumferential motion. The relative pathline or streamline of the liquid particle with the initial point ① on the plane of rotating disc are showed in Fig.3(b). The motion direction of the liquid particle changes from radial direction at section ① to circumferential direction at section ④, due to the increasing Coriolis acceleration.

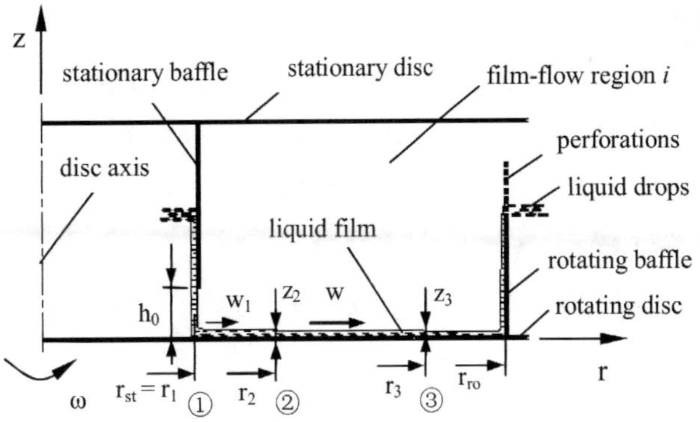

a) side view

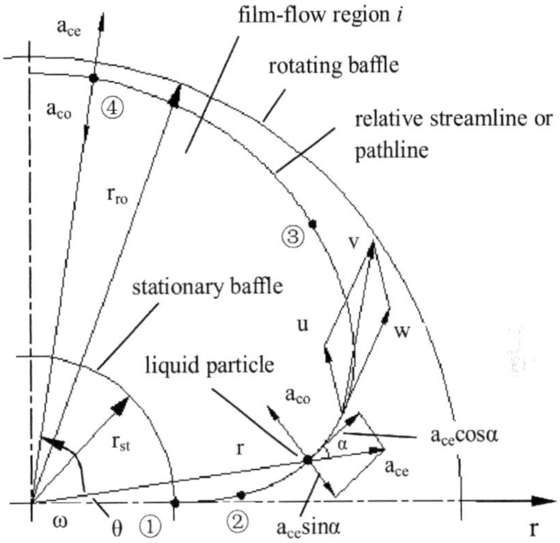

b) top view

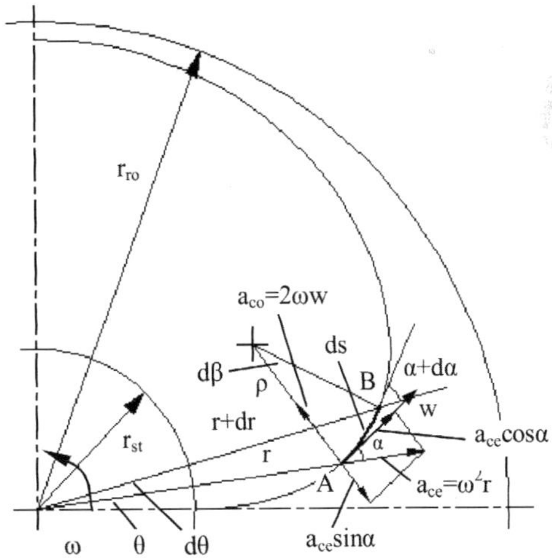

(c) small elemental section of *AB* with length *ds* along the streamline

Fig.3 Model of the ideal flow of liquid on the rotating disc between the stationary baffle and the rotating baffle

According to the assumption (3), the velocity, w_1, of the liquid onto the rotating disc falling from the bottom of the stationary baffles is

$$w_1 = \sqrt{2gh_0} \qquad (4)$$

And the velocity, w_1, is just the relative velocity at the initial point ① of the relative pathline or streamline on the rotating disc, because the velocity, w_1, is invariable when the liquid flows turn horizontal direction from vertical direction, as shown in Fig.3(a).

Writing the continuity equation from section ② to ③ in Fig.3(a) and (b), we have

$$Q_L = z_2 2\pi \cdot r_2 w_2 \cos\alpha_2 \cdot 3600 = z_3 2\pi \cdot r_3 w_3 \cos\alpha_3 \cdot 3600 \qquad (5)$$

The Bernoulli equation in rotating coordinate from section ② to ③ is given by Frank M.W.(2003) as

$$z_3 - z_2 = \frac{\omega^2}{2g}\left(r_3^2 - r_2^2\right) + \frac{w_2^2 - w_3^2}{2g} + \frac{p_2 - p_3}{\rho g} \qquad (6)$$

where $p_2 = p_3 = 0$, due to the case of liquid film flow with free liquid surface.

Substituting Eq.(5) into Eq.(6) yields:

$$w_3^3 - \left[2gz_2 + \omega^2\left(r_3^2 - r_2^2\right) + w_2^2\right]w_3 + 2g\frac{z_2 r_2 w_2 \cos\alpha_2}{r_3 \cos\alpha_3} = 0 \qquad (7)$$

From section ① to ③, the Eq.(7) become

$$w_3^3 - \left[2gz_1 + \omega^2\left(r_3^2 - r_1^2\right) + w_1^2\right]w_3 + 2g\frac{z_1 r_1 w_1 \cos\alpha_1}{r_3 \cos\alpha_3} = 0 \qquad (8)$$

The liquid falling from the stationary baffles flows to the radial direction at section ① in the film-flow region *i*, so the angle $\alpha_1=0$, as shown in Fig.3(b). The relative velocity, w_1, is obtained from Eq.(4), and the radius, r_1, is the radius of the stationary baffle, r_{st}. So the liquid film height, z_1, is written as

$$z_1 = \frac{Q_L/3600}{w_1 \cdot 2\pi \cdot r_{st}} \qquad (9)$$

The angle α_3 depends on r_3. And if α_3 is obtained, w_3 will be obtained from Eq.(8), so we can obtain the absolute velocity, v_3, as the resultant of the relative velocity, w_3, and the circumferential velocity, u_3, with $u_3=\omega r_3$. According to Euler's turbine equation(Frank M.W. 2003), the power, N'_{LDi}, delivered to the liquid in the film-flow region *i*, when $r_{st}= r_1$ and $r_{ro}= r_3$, is

$$\begin{aligned} N'_{LDi} &= \omega T = \rho_L\left(Q_L/3600\right)\cdot 0.5\cdot\left(v_3^2 - v_1^2 + u_3^2 - u_1^2 - \left(w_3^2 - w_1^2\right)\right) \\ &= \rho_L\left(Q_L/3600\right)\left(u_3^2 - u_1^2 + u_3 w_3 \sin\alpha_3\right) = \rho_L\left(Q_L/3600\right)\left(u_{ro}^2 - u_{st}^2 + u_{ro} w_{ro} \sin\alpha_{ro}\right) \end{aligned} \qquad (10)$$

where u_{ro}, u_{st}, w_{ro} and α_{ro} is equal to u_3, u_1, w_3 and α_3, respectively.

Because the circumferential velocity of the liquid falling from the stationary baffles is zero, the power required for the acceleration of the liquid to the circumferential velocity, ωr_{st}, is

$$N''_{LDi} = 0.5\rho_L\left(Q_L/3600\right)\cdot r_{st}^2\omega^2 \qquad (11)$$

So the whole power, N_{LDi}, delivered to the liquid in the film-flow region *i* is

$$\begin{aligned} N_{LDi} &= N'_{LDi} + N''_{LDi} = \rho_L\left(Q_L/3600\right)\left(u_{ro}^2 - u_{st}^2 + u_{ro} w_{ro} \sin\alpha_{ro}\right) + 0.5\rho_L\left(Q_L/3600\right)\cdot u_{st}^2 \\ &= \rho_L\left(Q_L/3600\right)\left(u_{ro}^2 - 0.5u_{st}^2 + u_{ro} w_{ro} \sin\alpha_{ro}\right) \end{aligned} \qquad (12)$$

The whole rotor has m film-flow regions, so the overall power is

$$N_{LD} = \sum_{i=1}^{m} N_{LDi} \qquad (13)$$

The Eq.(13) is just the theoretical equation of calculating the power requirement, N_{LD}, of the ideal liquid flow in the RZB rotor without gas flow.

Now a model is developed for the function of α_{ro} as r_{ro}. The position of the point *A* on the streamline is determined by the polar coordinates of parameters of radius, *r*, and radian, θ, as shown in Fig.3(c). So the motion equation of the point *A* is

$$\begin{cases} r = r(t) \\ \theta = \theta(t) \end{cases} \qquad (14)$$

The centrifugal acceleration of the point *A* is $\omega^2 r$ with the relative velocity, *w*. A small elemental section of *AB* with length *ds* along the streamline is taken in Fig.3(c), the position of the point *B* on the streamline by the polar coordinates of parameters of radius, *r+dr*, and radian, $\theta+d\theta$, from which we obtain the equation of $ds\cdot\cos\alpha = dr$.

And the equation of motion for the liquid particle and the geometric equation for the elemental section *AB* may be written as

$$\frac{d\left(\dfrac{ds}{dt}\right)}{dt} = \frac{d\left(\dfrac{dr/\cos\alpha}{dt}\right)}{dt} = a_\tau = a_{ce}\cos\alpha = \omega^2 r\cos\alpha \qquad (15a)$$

$$\rho\cdot d\beta = ds \qquad (15b)$$

where ρ is the curvature radius, and the ρ of point *A* is equal to that of *B* due to the small elemental section of *AB*, from which $d\beta = d\alpha$ .Integrating Eq.(15a) gives

$$ds = \left(\omega^2 \int r\cos\alpha dt\right)dt \qquad (16)$$

The curvature radius ρ is given as

$$\rho = \frac{w^2}{a_n} = \frac{w^2}{a_{co} - a_{ce}\sin\alpha} = \frac{w^2}{2\omega \cdot w - \omega^2 r \sin\alpha} \qquad (17)$$

Substituting $d\beta = d\alpha$, Eq.(17) and $w = ds/dt$ into Eq.(15b) yields:

$$\frac{(ds/dt)^2 d\alpha}{2\omega \cdot (ds/dt) - \omega^2 r \sin\alpha} = ds \qquad (18)$$

Substituting Eq.(16) into Eq.(18) yields:

$$\frac{\left(\omega^2 \int r\cos\alpha dt\right)^2 d\alpha}{2\omega \cdot \left(\omega^2 \int r\cos\alpha dt\right) - \omega^2 r \sin\alpha} = \left(\omega^2 \int r\cos\alpha dt\right) dt \qquad (19)$$

which reduces to

$$\left(\int r\cos\alpha dt\right) d\alpha = \left[2\omega\left(\int r\cos\alpha dt\right) - r\sin\alpha\right] \cdot dt \qquad (20)$$

Differentiation on the two sides of Eq.(20) gives

$$(r\cos\alpha dt) d\alpha = [2\omega(r\cos\alpha dt) - d(r\sin\alpha)] \cdot dt \qquad (21)$$

which reduces to

$$2\frac{d\sin\alpha}{dt} + \sin\alpha \cdot \frac{d\ln r}{dt} = 2\omega \cdot \cos\alpha \qquad (22)$$

Applying the dimensionless radius $\bar{r} = \frac{r}{r_{st}}$ ($\bar{r} \geq 1$) and the dimensionless time $\bar{\tau} = \frac{t}{\pi/(2\omega)}$, Eq.(22) is non-dimensionalized as

$$2\frac{d\sin\alpha}{d\bar{\tau}} + \sin\alpha \cdot \frac{d\ln\bar{r}}{d\bar{\tau}} = 2\omega \cdot \cos\alpha \cdot \left(\frac{\pi}{2\omega}\right) \qquad (23)$$

where the term $\sin\alpha \cdot \frac{d\ln\bar{r}}{d\bar{\tau}}$ has been omitted because the term is an order of magnitude small than the term $2\frac{d\sin\alpha}{d\bar{\tau}}$ when small $\bar{r}$ leads to small $\sin\alpha$, that is

$$\frac{d\sin\alpha}{d\bar{\tau}} = \omega \cdot \cos\alpha \cdot \left(\frac{\pi}{2\omega}\right) \qquad (24)$$

With the initial condition that α=0 at τ=0, Equation (24) can be solved as

$$\alpha = (\pi/2)\bar{\tau} = \omega \cdot t \qquad (25)$$

Now, we begin to solve the differential equation of Eq.(15a). Substituting Eq.(25) into Eq.(15a) yield:

$$\frac{d\left(\frac{1}{\cos(\omega t)}\frac{dr}{dt}\right)}{dt} = \omega^2 r\cos(\omega \cdot t) \qquad (26)$$

which reduces to

$$\frac{d^2 r}{dt^2} + \frac{dr}{dt} tg(\omega \cdot t)\omega = \omega^2 r\cos^2(\omega \cdot t) \qquad (27)$$

We substitute $\frac{dr}{dt} = \cos(\omega t)\int_0^t \omega^2 r\cos(\omega t)dt$ obtained from Eq.(26) into Eq.(27), thus

$$\frac{d^2 r}{dt^2} + \int_0^t \omega^2 r\cos(\omega t)dt \cdot \sin(\omega \cdot t)\omega = \omega^2 r\cos^2(\omega \cdot t) \qquad (28)$$

Using the dimensionless radius $\bar{r} = \frac{r}{r_{st}}$ and the dimensionless time $\bar{\tau} = \frac{t}{\pi/(2\omega)}$, Eq.(28) is non-dimensionalized as

$$\frac{d^2\bar{r}}{d\bar{\tau}^2} + \int_0^{\bar{\tau}} \bar{r}\cos\left(\frac{\pi}{2}\bar{\tau}\right)d\bar{\tau} \cdot \sin\left(\frac{\pi}{2}\bar{\tau}\right) \cdot \frac{\pi^3}{8} = \bar{r}\cos^2\left(\frac{\pi}{2}\bar{\tau}\right)\frac{\pi^2}{4} \qquad (29)$$

The function of $\bar{r}$ as $\bar{\tau}$ will be obtained when Eq.(29) was solved. The polynomial approximation method (the method is left out) gives the solution of Eq.(29) as

$$\bar{r}(\bar{\tau}) = (-1.740 + 0.5\pi Lo)\bar{\tau}^3 + (2.610 - \pi Lo)\bar{\tau}^2 + 0.5\pi Lo\bar{\tau} + 1 \qquad (30)$$

where *Lo* is a new dimensionless number, defined as $Lo = w_1/(\omega \cdot r_{st})$, which indicates the ratio of Coriolis acceleration to centrifugal acceleration at section ①. And the Eq.(30) is applicable for smaller $\bar{r}$.

3. Experiment

Experiments were carried out with air-water system on a RZB that is shown schematically in Fig.4. The air was introduced into the RZB tangentially by a blower, flowing inwards. And the water entered the RZB at the eye of the RZB rotor using a high tank, flowing outwards. The motor drove the rotor of the RZB to rotate with the pulley. The rotational speed of the motor was varied using a frequency modulator.

The casing of the RZB was of 320mm diameter and 800mm height. The rotor of the RZB with 600mm diameter and 80mm height is composed of six rotating baffles and six stationary baffles. The perforations are made with 1mm diameter of holes and 2mm distance between holes on the upper portion with 33mm distance from the bottom of the rotating baffles. Table 1 presents the diameters of the rotating and stationary baffles, in which film-flow region 2~6 is between the inner stationary baffle and the adjacent outer rotating baffle, and region 1 is between liquid inlet pipe with 17mm inner diameter and the adjacent rotating baffle with 250mm diameter. The average radius of the rotating baffles, $r_{ro,avg}$, is 437.3/2=218.7mm with the average moist periphery of liquid film, $l_{avg}=2\pi r_{ro,avg}=1373.4$mm. And the average gap between inner rotating baffle and outer stationary baffle in mass-transfer region, $\triangle r_{avg}$, is 20.4mm with the average area of annular mass-transfer region, A_{avg}, of 27957.1mm^2 in Table 1. The gap between the rotating baffles and the upper disc is 10mm, and the gap between stationary baffles and the lower disc is 20mm.

The liquid rate was measured with a rotameter, and the gas rate was measured with an orifice flowmeter. The rotational speed of the rotor was measured by a laser tachometer. Liquid rate, Q_L, ranged from 0m^3/h to 1.35m^3/h; and gas rate, Q_G, ranged from 0m^3/h to 600m^3/h. The rotational speed, n, ranged from 600rpm to 1400 rpm. The input power, N_{in}, of the motor was measured by an electric meter, from which the power requirement for RZB can be obtained.

The shaft power, N_B, can be obtained from the input power, N_{in}, of the motor through the motor efficiency, η_m, of 95% and the transmission efficiency, η_t, of 92%, which is given as

$$N_B = N_{in} \cdot \eta_m \cdot \eta_t \qquad (31)$$

The N_L in Eq.(1) as a function of rotational speed, gas rate and liquid rate may be obtained from N_B with various rotational speed, gas rate and liquid rate minus N_B with the same rotational speed and gas rate without liquid rate. The N_G as a function of rotational speed and gas rate may be obtained from N_B with various rotational speed and gas rate without liquid rate minus N_B with the same rotational speed without gas rate and liquid rate. The sum of N_w and N_{bf} at various rotational speed may be obtained as N_B without gas rate and liquid rate.

The N_{LR} in Eq.(2) and Eq.(3) as a function of rotational speed and liquid rate may be obtained from N_B with various rotational speed and liquid rate without gas rate minus N_B with the same rotational speed without liquid rate and gas rate.

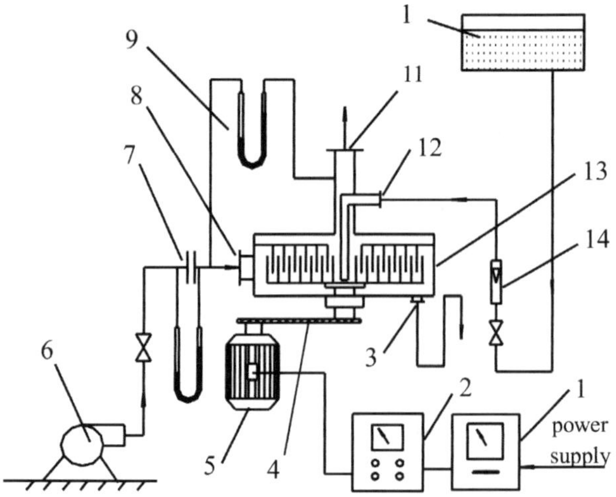

Fig.4. Schematic diagram of the experimental apparatus
1-electric meter; 2-frequency modulator; 3-liquid outlet; 4-pulley; 5-motor; 6-blower; 7-orifice flowmeter; 8-gas inlet; 9-U-tube differential pressure gauge; 10-high tank; 11-gas outlet; 12-liquid inlet; 13-RZB; 14-rotameter

Table 1 Dimensions of rotating and stationary baffles

No.	diameter of stationary baffle (mm)	diameter of rotating baffle (mm)	film-flow region no.	mass-transfer region	
				$\triangle$r (mm)	A (mm^2)
1		250	Region 1	27	23484.1
2	304		Region 2		
3		341		24	27506.4
4	389		Region 3		
5		419		21	29013.6
6	461		Region 4		
7		484		17.5	27557.4
8	519		Region 5		
9		541		17	29786.0
10	575		Region 6		
11		589		16	30395.2
12	621				
average	478.2	437.3		20.4	27957.1

3. Result

3.1 Comparison of N_{LD} and N_{LR} at various liquid rate and rotational speed

In Table 2, N_{LD} is calculated from α obtained in Eq.(30) which is applicable due to smaller $\bar{r}$ equal to the ration of r_{ro} and r_{st}. The comparison of N_{LD} and N_{LR} at various liquid rate and rotational speed is also shown in Table 2. Using the nonlinear regression analysis, experimental data allows us to obtain the correlation of K_2 in Eq.(3) as

$$K_2 = 0.03663 \mathrm{Re}_L^{\ 0.02137} \mathrm{Re}_\omega^{\ 0.1866} \qquad (32)$$

where Re_L is Reynolds number of the liquid film flow $(=4(Q_L/3600)\rho_L/l_{avg}\mu_L)$, and Re_ω is the rotational Reynolds number$(=\omega r_{ro,avg}^2 \rho_L/\mu_L)$. With a comparison of predicted and measured K_2, 80% of the points have a deviation of less than 10% and that the average deviation is about 4.86% with the greatest deviation below 15%. The value of K_2, ranging from 0.586 to 0.820, denotes that the power for the real liquid flow, N_{LR}, is less than that for the ideal liquid flow, N_{LD}. The circumferential velocity of the liquid falling on the rotating disc is zero, so it is impossible that the circumferential velocity of the liquid film reaches up to the circumferential velocity of the rotating disc for such a short time during which the liquid film rapidly flows through the annular region. Therefore, the value of N_{LR} reduces greatly in spite of the

Table 2 Value of N_{LD} in each film-flow region and the comparison of N_{LD} and N_{LR}

rotational speed (rpm)	Liquid rate (m^3/h)	Region 1		Region 2			Region 3			Region 4			Region 5			Region 6					
				$\bar{r}_2$=1.122			$\bar{r}_3$=1.077			$\bar{r}_4$=1.050			$\bar{r}_5$=1.042			$\bar{r}_6$=1.024					
		α	N_{LD1} (W)	$\bar{\tau}$	α	N_{LD2} (W)	$\bar{\tau}$	α	N_{LD3} (W)	$\bar{\tau}$	α	N_{LD4} (W)	$\bar{\tau}$	α	N_{LD5} (W)	$\bar{\tau}$	α	N_{LD6} (W)	N_{LD} (W)	N_{LR} (W)	K_2
1330	0.4	π/2	67.14	0.224	0.352	47.51	0.175	0.275	63.31	0.139	0.218	77.22	0.128	0.200	93.91	0.0948	0.149	103.8	452.86	344.80	0.761
	0.6	π/2	100.7	0.224	0.352	71.27	0.175	0.275	94.97	0.139	0.218	115.8	0.128	0.200	140.9	0.0948	0.149	155.6	679.32	522.56	0.769
	0.8	π/2	134.4	0.224	0.352	95.03	0.175	0.275	126.6	0.139	0.218	154.4	0.128	0.200	187.8	0.0948	0.149	207.5	905.81	742.71	0.820
	1.0	π/2	168.0	0.224	0.352	118.8	0.175	0.275	158.3	0.139	0.218	193.1	0.128	0.200	234.8	0.0948	0.149	259.4	1132.3	906.16	0.800
	1.24	π/2	208.5	0.224	0.352	147.3	0.175	0.275	196.3	0.139	0.218	239.4	0.128	0.200	291.1	0.0948	0.149	321.7	1404.2	1145.1	0.815
1200	0.4	π/2	54.66	0.223	0.350	38.64	0.174	0.274	51.52	0.138	0.217	63.57	0.127	0.200	76.43	0.0943	0.148	84.46	369.28	253.56	0.687
	0.6	π/2	82.02	0.223	0.350	57.97	0.174	0.274	77.28	0.138	0.217	94.26	0.127	0.200	114.6	0.0943	0.148	126.7	552.85	428.76	0.776
	0.8	π/2	109.4	0.223	0.350	77.29	0.174	0.274	103.0	0.138	0.217	125.7	0.127	0.200	152.9	0.0943	0.148	168.9	737.18	580.96	0.788
	1.0	π/2	136.8	0.223	0.350	96.61	0.174	0.274	128.8	0.138	0.217	157.1	0.127	0.200	191.1	0.0943	0.148	211.1	921.55	726.72	0.789
	1.24	π/2	169.8	0.223	0.350	119.8	0.174	0.274	159.7	0.138	0.217	194.8	0.127	0.200	236.9	0.0943	0.148	261.8	1142.9	908.61	0.795
1000	0.4	π/2	37.99	0.221	0.346	26.80	0.173	0.271	35.75	0.137	0.215	43.63	0.126	0.198	53.06	0.0932	0.146	58.65	255.88	167.15	0.653
	0.6	π/2	57.01	0.221	0.346	40.20	0.173	0.271	53.63	0.137	0.215	65.44	0.126	0.198	79.59	0.0932	0.146	87.98	383.85	282.44	0.736
	0.8	π/2	76.06	0.221	0.346	53.60	0.173	0.271	71.50	0.137	0.215	87.26	0.126	0.198	106.1	0.0932	0.146	117.3	511.84	397.39	0.776
	1.0	π/2	95.15	0.221	0.346	67.00	0.173	0.271	89.38	0.137	0.215	109.1	0.126	0.198	132.7	0.0932	0.146	146.6	639.88	498.89	0.780
	1.24	π/2	118.1	0.221	0.346	83.08	0.173	0.271	110.8	0.137	0.215	135.2	0.126	0.198	164.5	0.0932	0.146	181.8	793.59	611.53	0.771
800	0.4	π/2	24.28	0.217	0.341	17.08	0.170	0.267	22.80	0.135	0.211	27.84	0.124	0.195	33.87	0.0916	0.144	37.45	163.32	107.89	0.661
	0.6	π/2	36.45	0.217	0.341	25.62	0.170	0.267	34.20	0.135	0.211	41.75	0.124	0.195	50.81	0.0916	0.144	56.18	245.01	179.25	0.732
	0.8	π/2	48.64	0.217	0.341	34.16	0.170	0.267	45.60	0.135	0.211	55.67	0.124	0.195	67.75	0.0916	0.144	74.91	326.73	251.71	0.770
	1.0	π/2	60.88	0.217	0.341	42.70	0.170	0.267	57.00	0.135	0.211	69.59	0.124	0.195	84.68	0.0916	0.144	93.63	408.49	303.44	0.743
	1.24	π/2	75.63	0.217	0.341	52.95	0.170	0.267	70.68	0.135	0.211	86.29	0.124	0.195	105.0	0.0916	0.144	116.1	506.66	380.66	0.751
600	0.4	π/2	13.68	0.211	0.332	9.57	0.166	0.260	21.09	0.131	0.206	15.64	0.121	0.190	19.03	0.0890	0.140	21.06	100.06	58.652	0.586
	0.6	π/2	20.54	0.211	0.332	14.36	0.166	0.260	27.48	0.131	0.206	23.45	0.121	0.190	28.55	0.0890	0.140	31.58	145.97	95.392	0.654
	0.8	π/2	27.43	0.211	0.332	19.15	0.166	0.260	34.71	0.131	0.206	31.27	0.121	0.190	38.07	0.0890	0.140	42.11	192.74	134.04	0.695
	1.0	π/2	34.36	0.211	0.332	23.93	0.166	0.260	19.56	0.131	0.206	39.09	0.121	0.190	47.58	0.0890	0.140	52.64	217.17	172.11	0.793
	1.24	π/2	42.75	0.211	0.332	29.68	0.166	0.260	26.41	0.131	0.206	48.47	0.121	0.190	59.00	0.0890	0.140	65.28	271.58	199.10	0.796

increase of power requirement due to the friction losses for the real liquid.

3.2 Comparison of N_{LR} and N_L at various rotational speed, liquid rate and gas rate

When the gas enters the rotor of RZB, the power requirement associated with the flow of liquid is N_L to which N_{LR} is converted. The effects of the gas rate on the N_L are: (1) the gas flows on the surface of the liquid film with higher circumferential velocity of the gas than that of the liquid film in the film-flow region due to the conversation of angular momentum of the inward-flow gas, which pushes the liquid film to flow, so N_L reduces; (2)some of fine liquid droplets leaving from the rotating baffles are entrained upwards and inwards by the gas in the mass-transfer region, as shown in Fig.2, and is caught up by the liquid film on the rotating disc, which increases the liquid rate, Q_L, of N_L, so N_L increases; and (3) the gas tears the falling liquid from the stationary baffles to droplets, and the droplets fly inwards, as shown in Fig.2, being equivalent to increase the length of the liquid film flow between the stationary and rotating baffles in the film-flow region, which increases N_L.

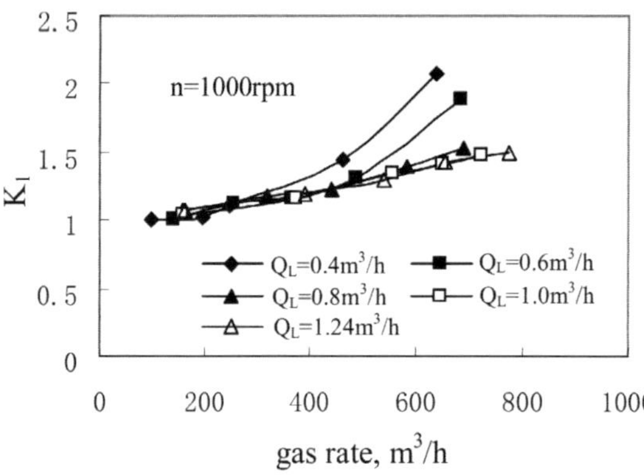

Fig.5 K_I at various gas rate and liquid rate with rotational speed of 1000rpm

It can be seen from Fig.5 that K_I, as the ratio of the measured N_L and N_{LR} in Eq.(2), increases with increasing gas rate. The K_I is less than 1.0 at the small gas rate, because effect (2) of the gas pushing the liquid film to flow is more dominant than effect (1) and (3), leading to N_L less than N_{LR}. And the K_I is more than 1.0 at the great gas rate, because effect (1) and (3) become more dominant.

We introduce the gas Reynolds number, $Re_G(=4R_h u_G \rho_G/\mu_G)$, to analyze the effect of the gas rate on the N_L. The effect of entrainment of the liquid droplets in the mass-transfer region on the N_L is greatest in the three effects, so the gas velocity, u_G, of Re_G is based on the annular mass-transfer area, A_{avg}, and the hydrodynamic radius, R_h, is also from the cross section of annular gap in the mass-transfer region. Using the nonlinear regression analysis, experimental data allows us to obtain the correlation of K_I in Eq.(2) as

$$\begin{cases} K_1 = 1 & at\ Q_G = 0 \\ K_1 = 6.5480\,Re_L^{-0.10735}\,Re_\omega^{-0.23693}\,Re_G^{0.29279} & at\ Q_G > 0 \end{cases} \qquad (33)$$

where Re_G is $4R_h u_G \rho_G/\mu_G$ with $R_h = \Delta r_{avg}/2$ and $u_G = (Q_G/3600)/A_{avg}$. With a comparison of predicted and measured K_I, 80% of the points have a deviation of less than 10% and the average deviation is about 6.06% with the greatest deviation below 30%. The value of K_I ranges from 0.929 to 1.694.

It can be seen from Eq.(33) that K_I is reduced with increasing Re_ω and Re_L. The velocity of the liquid droplets leaving from the rotating baffles increases rapidly with increasing Re_ω, which reduces the entrainment of the liquid droplets by the gas and then reduces N_L and K_I. And the size of the liquid droplets leaving from the rotating baffles increases because the coalescence between the liquid droplets enhances with increasing Re_L, which also reduces the entrainment by the gas and then reduces N_L and K_I.

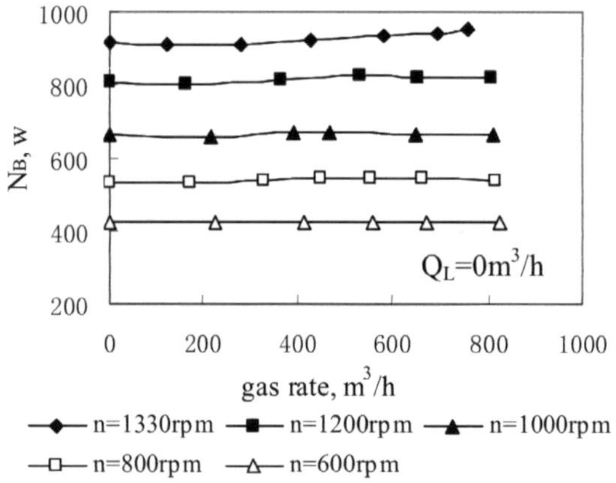

Fig.6 Shaft power, N_B, at various rotational speed and gas rate without liquid rate

3.3 Effect of N_G

The shaft power, N_B, at various rotational speed and gas rate without liquid rate is shown in Fig.6. The N_G is the value difference between the N_B with gas rate and without gas rate at the same rotational speed. And the N_G approximates to zero, because the N_B changes very little when the gas rate increases. Therefore, the N_G can be negligible because it only occupies 0.4%~1.4% of the N_B, which is in agreement with the conclusion of Keyvani(1989).

3.4 Effect of N_W and N_{bf}

The sum of N_W and N_{bf} is the shaft power, N_B, at various rotational speed without gas rate and liquid rate in Fig.7. Because N_W is proportional to $\omega^{1.5}$ and N_{bf} is proportional to ω (Keyvani 1989), experimental data allows us to obtain the correlation as

$$N_W + N_{bf} = 0.06358\omega^{1.5} + 5.74827\omega^{1.0} = 6.07839\times10^{-9}\,Re_\omega^{1.5} + 1.20182\times10^{-4}\,Re_\omega^{1.0} \qquad (34)$$

From Eq.(34), N_{bf} is greater, up to about 10 times N_W. The N_{bf} includes power consumed in the bearing friction and the dynamic seal friction in Fig.1. The seal system of RZB is complicated for acquiring the good seal which lead to the greater N_{bf}. The sum of N_W and N_{bf} occupies the high percentage of the N_B due to the greater N_{bf}.

4. Conclusion

(1)The relative streamline or pathline of the liquid particle moving on the rotating disc is proposed using an approximate solution of the mathematic differential equation of the liquid particle on which the centrifugal and Coriolis acceleration are exerted.

(2) The power requirement, N_{LD}, for the ideal liquid flow in the RZB rotor without gas flow is calculated by a theoretical equation of Eq.(12). And the effect of the liquid density, ρ_L, and the liquid rate, Q_L, on N_{LD} can be obtained from Eq.(12).

(3) The power requirement, N_{LR}, for the real liquid flow in the RZB rotor without gas flow is equal to N_{LD} multiplied by K_2. The K_2 is correlated by the experimental data, ranging from 0.586 to 0.820. The K_2 depends on the liquid rate and rotational speed, due to zero of the circumferential velocity of the liquid falling on the rotating disc and the difference of the friction losses between ideal and real liquid.

(4) The power requirement, N_L, for the real liquid flow in the RZB rotor with gas flow is equal to N_{LR} multiplied by K_1. And the K_1 is correlated by the experimental data, ranging from 0.929 to 1.694. The K_1 depends on the liquid rate, gas rate and rotational speed, due to three gas effects of propelling the liquid film flow by gas, liquid droplets entrainment by gas and the increase in the length of the liquid film flow by gas.

(5) The shaft power, N_B, is the sum of N_L, N_W, N_{bf} and N_G. The N_G can be negligible because it only occupies 0.4%~1.4% of the N_B. The sum of N_W and N_{bf} is also correlated by the experimental data, so the shaft power, N_B, of the RZB is calculated using several above correlations.

Nomenclature

a_{co} — Coriolis acceleration, m/s^2
a_{ce} — centrifugal acceleration, m/s^2
a_n — normal acceleration, m/s^2
a_τ —tangential acceleration, m/s^2
A_{avg}— average area of annular mass-transfer region between inner rotating baffle and outer stationary baffle, m^2
h_0 — gap between the stationary baffles and the lower rotating disc, m
K_1 — coefficient for equation (2)
K_2 — coefficient for equation (3)
l_{avg} — moist periphery of liquid film, $=2\pi r_{ro,avg}$, m
Lo — dimensionless number indicating the ratio of Coriolis acceleration to centrifugal acceleration at section ①
m — number of film-flow regions in the whole rotor of the RZB
n — rotational speed, rpm
N_B— shaft power of the RZB, W
N_{in} — input power of the motor of the RZB, W
N_L— power requirement for the flow of liquid through the rotor including the liquid acceleration and friction losses, W
N_w — power consumed in overcoming the frictional windage drag of the rotor, W
N_{bf}— power required for the bearing friction, W
N_G— power requirement for the flow of gas through the rotor, W
N_{LR} — power requirement for the real viscous liquid flow without gas flow, W
N_{LD}— power requirement for the ideal liquid flow without gas flow, W
N_{LDi}— power requirement for the ideal liquid flow without gas flow in the film-flow region *i*, W
N'_{LDi}— power which is delivered to the ideal liquid flow in the film-flow region *i*, according to Euler's turbine equation, W
N''_{LDi}— power required for the acceleration of the liquid to the circumferential velocity, ωr_{st}, in the film-flow region *i*, W
p — liquid pressure, Pa; p_2, p_3, at section ②, ③, respectively
Q_L — liquid rate, m^3/h
Q_G — gas rate, m^3/h
r — radial distance from the rotating disc axis, m; r_1, r_2, r_3, at section ①, ②, ③, respectively
$\bar{r}$ — dimensionless radius;
r_{st} — radius of the stationary baffle, m
r_{ro} — radius of the rotating baffle, m
$\triangle r_{avg}$— average gap between inner rotating baffle and outer stationary baffle in mass-transfer region, m
$r_{ro,avg}$ — average radius of the rotating baffles, m
s — arc coordinate of the liquid particle on the pathline or streamline, m
t — time, s
T — torque about the rotating disc axis, Nm
u — circumferential velocity of liquid film $u=\omega r$, m/s; u_1, u_3, at section ①, ③, respectively
u_{ro} —circumferential velocity of liquid film at the radius of the rotating baffle, m/s
u_{st} —circumferential velocity of liquid film at the radius of the stationary baffle, m/s
v — absolute velocity of liquid film on the rotating disc, m/s; v_1, v_3, at section ①, ③, respectively
w — relative velocity of liquid film on the rotating disc, m/s; w_1, w_2, w_3, at section ①, ②, ③, respectively
$\bar{w}$ — dimensionless relative velocity;
w_1 — velocity of the liquid onto the rotating disc falling from the bottom of the stationary baffle (equal to the relative velocity of liquid film on the rotating disc at section ①), m/s

w_{ro} — relative velocity of liquid film at the radius of the rotating baffle, m/s

z — distance along the rotating disc axis, m; z_1, z_2, z_3, at section ①, ②, ③, respectively

Greek Symbols

ω — angular speed, rad/s

α — angle with tangential direction of the relative pathline to radial direction, rad; α_1, α_2, α_3, at section ①, ②, ③, respectively

α_{ro} — angle with tangential direction of the relative pathline to radial direction at the rotating baffle, rad

θ — rotation angle about the rotating disc axis, rad

ρ — curvature radius, m

ρ_L — liquid density, kg/m^3

β — angle between the curvature radius of point A and B

$\bar{\tau}$ — dimensionless time;

η_m — motor efficiency

η_t — transmission efficiency

Reference

Basic, A. and Dudukovic, M.P. Liquid holdup in rotating packed beds: Examination of the film flow assumption, AICHE J, 1995, 41:301-316

Burns, J.R., Jamil, J.N. and Ramshaw, C. Process intensification: operating characteristics of rotating packed beds—determination of liquid hold-up for a high-voidage structured packing, Chem.Eng.Sci. 2000, 55, 2401-2415

Burns, J.R. and Ramshaw, C. Process intensification: visual study of liquid maldistribution in rotating packed beds, Chem.Eng.Sci. 1996, 51, 1347-1352

Chen, J.F., Shan,L., Guo,F. Synthesis of nano-fibers of aluminum hydroxide in novel rotating packed bed reactor, Chem.Eng.Sci. 2003, 58, 569-575

Chen, J.F., Wang, Y.H., Guo,F. Synthesis of nanoparticles with novel technology: High-gravity reactive precipitation, Ind.Eng.Chem.Res. 2000, 39. 948-954

Chen, Y.S., Lin, C.C., Liu, H.S. Mass Transfer in a Rotating Packed Bed with Viscous Newtonian and Non-Newtonian Fluid, Ind. Eng. Chem. Res. 2005, 44. 1043-1051

Das, A., Bhowal, A., Datta, S. Continuous Biosorption in Rotating Packed-Bed contactor, Ind.Eng.Chem.Res. 2008, 47. 4230-4235

Dhiman, S.K. Process intensification in a trickle-bed reactor: Experimental Studies, AICHE Journal, 51(12), December,2005, 3186-3192

Frank M.W., 2003, Fluid Mechanics, 5th edition (McGraw-Hill Education (Asia) Co. and Tsinghua University Press, Beijin, China).

Guo, F., Zheng, C., Guo, K. Hydrodynamics and mass transfer in crossflow rotating packed bed, Chem.Eng.Sci. 1997, 52, 3853-3859

Guo, K., Guo, F., Feng, Y. Synchronous visual and RTD study on liquid flow in rotating packed-bed contactor, Chem.Eng.Sci. 2000, 55, 1699-1706

Ji, J.B. Equipment of zigzag high-gravity rotating beds, China patent 01134321.4, Zhejiang University of Technology (China), 2004.

Ji, J.B., Xu, Z.C. Equipment of zigzag high-gravity rotating beds, U.S. Patent 7,344,126, 2008.

Keyvani, M. Operating Characteristics of rotating beds. Chemical Engineering Progress. 1989, 85: 48-52.

Kumar, M.P. and Rao, D.P. Studies on a high-gravity gas-liquid contactor, Ind.Eng.Chem.Res. 1990, 29, 917-920

Kelleher, T. and Fair, J.R. Distillation studies in a high-gravity contact, Ind.Eng.Chem.Res. 1996, 35, 4646-4655

Lin, C.C., Ho, T.J., Liu, W.T. Distillation in a Rotating Packed Bed, Journal of Chemical Engineering of Japan, 2002, 35(12), 1298-1304

Lin, C.C., Chen, Y.S. and Liu, H.S. Prediction of liquid holdup in countercurrent-flow rotating packed bed, Trans IChemE. April 2000, 78: 397-403

Lockett, M.J. Flooding of Rotating Structured Packing and its Application to Conventional Packed Columns. Trans IChemE. 1995, 73: 379-384

Mallinson, R.H., Ramshaw, C. Mass transfer Apparatus and process. European Patent 0053881, 1982.

Matsumoto, S., Saito, K. and Takashima, Y., 1973, The thickness of a viscous liquid film on a rotating disk. J.Chem. Eng. Japan, 6, 503-507.

Munjal, S., Dudukovic, M.P., Ramachandram, P. Mass-transfer in rotating packed bed. II. Experimental results and comparison with theory and gravity flow. Chem.Eng.Sci. 1989, 44, 2257

Niqam, K.D.P. Process intensification in a trickle-bed reactors, Chemical Engineering Science, 66(22), November, 2005

Ramshaw, C., Mallinson, R.H. Mass Transfer Process. U.S. Patent 4,283,255, 1981.

Rao, D.P., Bhowal, A., Goswami, P.S. Process Intensification in Rotating Packed Beds(HIGEE): An Appraisal. Ind.Eng.Chem.Res. 2004, 43(4): 1150-1162

Singh, S.P., Wilson, J.H., Counce, R.M. Removal of volatile organic compounds from groundwater using a rotary air stripper, Ind.Eng.Chem.Res. 1992, 31(2): 574-580

Wang, G.Q.; Xu, Z.C.; Yu, Y.L.; Ji, J.B. Performance of a rotating zigzag bed-A new HIGEE, Chemical Engineering and Processing: Process Intensification,2008, 47(12), November, 2131-2139

Yang, H.J., Chu,G.W., Xiang,Y. Characterization of micromixing efficiency in rotating packed beds by chemical methods, Chemical Engineering Journal, 121(2006) 147-152

Zheng, C. Pressure drop of centripetal gas flow through rotating beds. Ind.Eng.Chem.Res. 2000, 39(3): 829-834

Increasing Power Savings in Removal of Oil from Industrial Effluents Via Electrocoagulation

ABSTRACT

Electrode passivation raises the power consumption, is a concern for the longevity of the process and increases the operating cost, and hence, is the greatest operational issue with the electrocoagulation process. The objectives of the current investigation were to define the reactor configuration and optimum working conditions to maximize the amount of oil removal from an industrial effluent with minimum specific energy consumption. We show that power consumption per unit effluent treated can be greatly reduced by using a half-wave rectified alternating-current supply instead of a full alternating-current supply. We also developed a theoretical model for predicting oil separated over time in an electrocoagulator. To this end, we modified the model of Matteson et al [1] by dividing the electrocoagulator space into two regions, viz. the bulk region, where hydrodynamic and/or Brownian forces are dominant, and the electrode region, where the electrophoretic force is dominant. The model was in a good agreement with the experimental data.

INTRODUCTION

Steel rolling mills discharge a metastable, spontaneously forming oil-in-water emulsion as its effluent. This effluent results from an emulsion which serves the dual purpose of lubrication and cooling in the metal working process of the reduction of the thickness of hot rolled steel sheets. The effluent streams of such industries are often difficult to treat because of their variety and complexity, ranging from simple free-floating oil to complex streams containing chemically or mechanically emulsified oils, fats, greases, soluble oils, dirt, suspended solids and toxic heavy metals or other hazardous substances. Many different techniques are available for the treatment of wastewater, including a variety of filters, chemical dosing, reverse osmosis, gravity separation, emulsion breaking, dissolved air flotation and ultra-filtration. Many of these techniques are either very pollutant specific or more expensive than dumping and using more water. The latter options are fast disappearing and new methods need to be sought.

The technique of electrocoagulation is a relatively new one and is thought to have been first proposed in 1889 in England (Matteson, 1995). In the electrocoagulation process for water treatment, Al or Fe electrodes are dissolved by electrolysis, forming a range of coagulant species and metal hydroxides, which destabilize and aggregate the suspended particles or droplets, and adsorb dissolved contaminants (e.g. dissolved organic matter) (Jiang, 2002). A variation of the electrocoagulation process is one in which the gas generated by the process captures the coagulated pollutants and floats them to the surface. The pollutants are then removed from the top by periodically adding water from the flush water reservoir, the amount of water added being about 5% of the volume of the water being processed (Robinson V.). Known as electroflocculation, it offers the possibility of a one step process, in which polluted water is pumped into a processing reactor, the pollutants are floated to the surface and the cleaned water is pumped out at the end of the process. Theoretically, there are no chemicals to be added and no filters that need cleaning.

The electrolytic addition of ions offers a number of advantages over their addition chemically. The electric field increases the probability of oppositely charged particles and ions in suspension to come into contact with each other. Therefore, the electrocoagulation process improves coagulation (Matteson, 1995; Mollah, 2001). There is no addition of anions meaning no increase in the salinity of the treated water. The system produces half to one third of the sludge. Greater activity of the electrolytically added aluminum ions means that less aluminum is required, and that this process can be used to treat a number of different pollutants that cannot be handled by chemical flocculants such as alum (aluminum sulphate). Electroflocculation is able to fulfill simultaneous coagulation and flotation (by tiny hydrogen bubbles produced at cathodes). Dosing control by adjusting current, no moving parts, and no external addition of chemicals, low power consumption and compact equipment are some of the additional benefits offered by electrocoagulation technique (Chen 2000; Eow 2002).

Electrode passivation, raising the power consumption, a concern for the longevity of the process and increasing the operating cost, is the greatest operational issue with the electrocoagulation process. The objectives of the investigation reported here were to define the reactor configuration and optimum working conditions to maximize the amount of oil removal with minimum specific energy consumption, and to find a solution for electrode passivation, using both experimental and theoretical approach.

Theory

Electrocoagulation involves the in-situ generation of ions by electrolysis. In general, dissolving metal anodes are used to continuously produce metal ions close to the anodic surface. The matteson's model was based on the electrophoretic movement of oil droplets / particles towards the anode, where they are discharged, and then circulate throughout the bulk where they encounter other discharged particles and coagulate (Matteson, 1995). Matteson et al. (1995) modeled the electrocoagulation process for a stirred suspension between two opposing flat steel wire mesh electrodes. They defined the loss of particles due to coagulation as the second order relation, given by

$$\frac{dC_T}{dt} = k_1 C_T^2 \left(1 - e^{-k_2 t}\right)^2 \tag{1}$$

Where, C_T is number concentration of suspension particles.

Analytical solution to the equation 3.3.1 is given by

$$C_n = \frac{C_T}{C_0} = \left[1 + k_4 C_0 t + \frac{k_4 C_0}{k_1}\left(-1.5 + 2e^{-k_1 t} - 0.5e^{-2k_1 t}\right)\right] \tag{2}$$

Where, C_n is fraction of initial particles left in the suspension, k_4 is 'Fast' Brownian coagulation constant, and, k_1 is electrophoretic migration rate constant.

In this work, we present a more rigorous approach to the electrocoagulation process, wherein separate mass balance equations were made for charged, uncharged, and electrode oil droplets / particles, and the cations. The movement of electrochemically generated cations in to the bulk, and subsequent discharging of charged droplets by a second order reaction is also considered.

The assumptions made to model the electrocoagulation process are summarized as follows. (1) All the drops were assumed to be spherical and of the same size. (2) Coalescence involving only two drops was considered. (3) The electrocoagulator space was divided into two regions: (a) The bulk region, where the hydrodynamic and/or Brownian forces are dominant. In this region, the oil drops are removed from the system by Brownian and shear induced collisions of uncharged particles with charged and other uncharged particles. The coagulation was assumed to be 'fast', i.e. every collision involving the uncharged particles leads to coalescence., (b) The region between the electrodes, where the electrophoretic force is dominant. Here the charged oil particles get discharged at the oppositely charged electrode, and a fraction of them move in to the bulk of suspension. In the boundary layer along the electrodes shear effects can be neglected, and hence the particles coalesce by only 'fast' Brownian coagulation.

Mass balance equation for cations

The cations diffuse into the bulk under the influence of the electrophoretic force. In the bulk, the discharge rate of charged oil particles is augmented by a second-order rate reaction between the cations and the charged oil droplets.

$$\frac{dC_i}{dt} = \left(\frac{dC_i}{dt}\right)_{electrophoretic\ migration} + \left(\frac{dC_i}{dt}\right)_{c-i\ collisions}$$

$$=> \quad V_{aq}\frac{dC_i}{dt} = uA_e\left(C_{eqb} - C_i\right) - k_6 V_{aq} C_i C_c \qquad (3)$$

Where, V_{aq} is volume of the aqueous phase, C_i is molar concentration of cations in the bulk, u is electrophoretic velocity given by equation 3.1.4, A_e is facial area of collecting electrode, C_{eq} is equilibrium (hydrolysis) molar concentration of cations, assuming cations are hydrolyzed as soon as they leave the electrode., C_{oh} is hydroxide ion concentration in the emulsion, k_6 is lumped parameter for the second order rate constant, inclusive of the effect of electrode area, electrode passivation, electrode geometry and material, and applied voltage, and, C_c is number concentration of charged particles i.e. number of particles per unit suspension volume (V_s)

The electrophoretic velocity of charged particles, u, towards the collector electrode may be expressed as the henry equation for non-conducting particles (Matteson, 1995):

$$u = \frac{E\xi\varepsilon}{1.5\mu} f(\kappa\alpha) \tag{3.1.4}$$

where

E is electric field strength between the electrodes,

ξ is the zeta potential,

ε is dielectric constant of continous phase,

μ is viscosity of continous phase.

$f(\kappa\alpha) = 1.5 \qquad for \qquad \kappa\alpha = 10.$

Mass balance equation for charged particles

The negatively charged oil particles are lost from the system in three ways- (1) Discharging by collisions with cations in the bulk region, (2) Discharging of particles by migration to the anode, and subsequent neutralization, and (3) Collisions with uncharged particles in the bulk region.

The discharge rate by collisions with cations is accounted for by the second order reaction term,

$$\left(\frac{dC_c}{dt}\right)_{c-i\ colisions} = -\left(\frac{dC_u}{dt}\right)_{c-i\ colisions} = -N_{av}k_6C_iC_c \tag{5}$$

The right hand side is multiplied by N_{av} (Avogadro's number) to convert the molar cation concentration in to the number concentration. The rate of migration of charged particles to the anode is given by (Matteson, 1995),

$$\frac{dC_c}{C_c} = -\frac{uA_e}{V_s}dt = -k_1dt \tag{6}$$

Assuming that all the particles reaching the anode are discharged immediately, the rate of migration of charged particles to the anode will be equal to the rate of generation of uncharged particles at the anode. This assumption gains weight from the fact that cation concentration at the anode, as given by Faraday's law, is much greater than the cation concentration in the bulk (C_i).

$$\left(\frac{dC_c}{dt}\right)_{migration} = -\left(\frac{dC_e}{dt}\right)_{anode} = -\frac{uA_e}{V_s}C_c = -k_1 dt \tag{7}$$

Where, C_e is number concentration of particles in the electrode boundary layer (based on V_s).

The Brownian and shear (BS) induced collisions between charged and uncharged particles in the bulk lead to creaming, and as a result, the particles are lost from the system. Freidlander has considered simultaneous BS coagulation to be additive, and shown good agreement with experiments. Hence, the simultaneous BS coagulation rate can be given by (Probstein, 1994),

$$\frac{dn}{dt} = -\left(\frac{4}{3}\frac{kT}{\mu} + \frac{16}{3}a^3 w\right)n^2 = -k_2 n^2 \tag{8}$$

where, w is shear rate, s^{-1}

Using (10), the rate of loss of oil droplets by BS induced collisions is given by,

$$\left(\frac{dC_c}{dt}\right)_{c-u\ coalescence} = \left(\frac{dC_u}{dt}\right)_{c-u\ coalescence} = -0.5k_2 C_c C_u \tag{9}$$

where, C_u is number concentration of uncharged particles.

The total rate of loss of charged particles from the system is given by adding the contributions of equations (6), (7), and (9),

$$\frac{dC_c}{dt} = -N_{av}k_6 C_i C_c - k_1 C_c - 0.5k_2 C_c C_u \tag{10}$$

Where, C_0 is initial number concentration of oil particles.

Mass balance equation for electrode particles

The charged particles migrate to the anode, get discharged and a fraction of the discharged particles bounce back into the bulk. The shear effects can be neglected in the electrode boundary layer. Hence, the particle loss at the electrodes occurs by 'fast' Brownian coagulation, and is given by (Probstein, 1994),

$$\frac{dn}{dt} = -8\pi DRn^2 = -\frac{4}{3}\frac{kT}{\mu}n^2 = -k_4 n^2 \tag{11}$$

The total rate of loss of charged particles from the system is given by adding the contributions of equations (7), (11), and subtracting the rate at which particles are bouncing back into the bulk.

$$\frac{dC_e}{dt} = \left(\frac{dC_e}{dt}\right)_{anode} + \left(\frac{dC_e}{dt}\right)_{bounce} + \left(\frac{dC_e}{dt}\right)_{e-e\ coalescence}$$
$$= k_1 C_c - k_3 C_e - k_4 C_e^2 \tag{12}$$

where, k_3 (= αk_1) is a constant to account for the rate at which particles are bouncing back into the bulk region, and, α is fraction of particles bouncing back into the bulk region.

Mass balance equation for uncharged particles

In addition to loss of uncharged particles by the u–c particle coalescence in the bulk, the u–u particle collisions in the bulk and the electrode region also lead to creaming of particles. The rate of change of uncharged particles in the system is given by their loss by creaming, and their generation by discharging of the charged particles.

$$\frac{dC_u}{dt} = -\left(\frac{dC_e}{dt}\right)_{bounce} + \left(\frac{dC_u}{dt}\right)_{c-i\ collisions} + \left(\frac{dC_u}{dt}\right)_{c-u\ coalescence} + \left(\frac{dC_u}{dt}\right)_{u-u\ coalescence}$$
$$= k_3 C_e + N_{av} k_6 C_i C_c - 0.5 k_2 C_c C_u - k_2 C_u^2 \tag{13}$$

Mass balance equation for creaming i.e. for the total loss of particles from the system

The particles are lost from the system by the c-u and u-u coalescence in the bulk region, and the e-e coalescence in the electrode boundary layer. The rate equation for creaming is given by,

$$\frac{dC_{cr}}{dt} = k_2 C_c C_u + k_2 C_u^2 + k_4 C_e^2 \tag{14}$$

or alternatively,

$$C_{cr}/C_0 = 1-(C_{cn}+C_{un}+C_{en}+C_{in}) \tag{15}$$

$$C_n = 1 - \frac{C_{cr}}{C_0} \tag{16}$$

where, C_n is fraction of initial particles left in the suspension.

If it is assumed that all the particles are uncharged at t = 0, the rate equation for creaming will be simply given by

$$\frac{dC_b}{dt} = -k_2 C_b^2 \tag{16}$$

Where, C_b is total particle concentration at any time, t.

The solution for equation 3.3.15 is given by

$$C_{bn} = \frac{C_b}{C_0} = \frac{1}{1 + k_2 C_0 t} \tag{17}$$

Where, C_{bn}: Fraction of initial particles left in the suspension if it is assumed that all the particles are uncharged at t = 0.

The set of coupled equations 10,12–14 was solved numerically using Runge-Kutta 4 scheme

Experimental procedure

Figure 1 shows the experimental apparatus with the electrode assembly. It was an electrocoalescer consisting of three concentric cylindrical electrodes of different diameters and heights (First cylinder: Diameter=14 cm, height=16 cm; Second cylinder: Diameter 12=cm, height=11 cm; Third cylinder: Diameter=10 cm, height=6 cm). The electrodes used in the experiments were aluminum plate electrodes (APE), and stainless steel plate electrodes (SSPE). The electrodes, mounted vertically on a sieved perspex disc at the top, were suspended inside a 5-litre perspex cylindrical reactor or electrocoalescer. The reactor was provided with an inlet at the top and an outlet at the bottom, making it suitable for use in both batch and continuous operations. A magnetic stirrer was used to agitate the contents of the reactor.

The reactor with the suspended electrode assembly was filled with 4.5-litres of emulsion effluent and kept on a magnetic stirrer. Alternate cylindrical electrodes were connected across an a.c., or two different unidirectional supplies. An ammeter was connected in series with the electrodes and a voltmeter was connected in parallel across the electrodes. A dimmerstat was used to provide the required a.c. voltage. For a given batch, 1-5 ml samples were withdrawn from the outlet at the bottom and their transmittance was measured at a known dilution using a spectrophotometer. The transmittance versus time data for each electrocoalescence run were converted to oil concentration versus time data using the spectrophotometer's calibration curve.

Results and Discussions

Electrode passivation and pitting

During the electrocoagualtion process, an inert coating of aluminum oxide develops on the surface and passivates the APE. Time and specific energy consumption (SEC) for 95 % oil removal with various types of current supplies are presented in table 1. In addition, table 2 gives the current densities and 95 % removal times with 25 V HWR a.c. across three different sets of APE. This difference in current-density values can be attributed to the additional resistance of the passivating aluminum-oxide layer. This results in a decrease in the rate of Al^{3+} dissolution, resulting in an increase in 95 %-oil removal times. The numbers in tables 1&2 clearly indicate the adverse effect of APE passivation for both a.c. and HWR a.c. supplies. This deterioration in APE-based electrocoagulator performance poses a serious design problem for effluent treatment. Another drawback of APE was that the electrodes showed pitting on use and ultimately dissolved away. This necessitated their replacement after 6-7 runs, as indicated in table 3. Both passivation and pitting put a limit on the electrode life, adding to the maintenance cost.

With SSPE, The 95 %-oil removal times for experimental run numbers 1 and 4 were respectively 25 and 22 minutes. The current density through the elctrocoalescer did not fall over an operational period of 13 experimental runs. Then there was a decrease through run 7 to 12. This anomaly can be attributed to improper cleaning of electrodes, as described below. Figure 3 shows this decrease in current density for runs 7 to 12. After run 12, the SSPE were taken out from the perspex top, and inspected. A thick organic deposit was observed on the outer surface of innermost electrode (anode) and the inner surface of outermost electrode (anode). The deposit could be easily scrubbed away, and the original condition of electrodes restored, as indicated by same value of current density at 10 V. HWR a.c. as was in run 7. The cleaning of SSPE electrodes can be done within the electrocoagulator, unlike the case with APE where it was essential to take out the electrodes to remove the passivating inorganic layer.

The SSPE are not actually passivated, which can be attributed to the non-inertness of ferrous oxide film. Further, SSPE did not show any signs of pitting after 12 runs. From these results, we can conclude that SSPE can be used over a long period of electrocoalescer operation without any deterioration in performance.

Current supply

The type of current supply had a very significant effect on the oil-removal time and power consumption. Three types of current supplies, viz. a.c., HWR a.c., and pulsed d.c. were used. From Table 1, the much greater efficacy of the unidirectional HWR a.c. supply is clearly demonstrated by the large decrease in the times for 95 % oil removal. In addition, only half the power will be consumed with HWR a.c. as compared with a.c. As a result there will be a huge saving in the energy consumption by switching over from a.c. to HWR a.c. supply, as indicated in table 5.1. Pulse d.c., another unidirectional power supply, was tested at 10 V for SSPE. In table 1, we compare pulse d.c. at two different frequencies, viz. 50 Hz and 200 Hz, against HWR a.c. Although there is not much difference in the oil-removal rates for the three different cases, the energy consumption was three times lower in the case of the pulse d.c. For APE, on switching over from a.c. to HWR a.c., the factors of reduction in 95 %-oil removal time and energy consumed were 2.5-8 and 2-9.5 respectively. Although the pulse d.c. decreased the energy consumption by a factor of 3 times, the high initial cost of making a pulse d.c. supply might make its use uneconomical. We may therefore conclude that the use of HWR a.c. supply is an important step towards making the electrocoagulation process for effluent treatment economical in practice.

The efficacy of unidirectional current supplies can be explained on the basis of electrophoretic force acting on the negatively charged oil droplets. Under the influence of unidirectional field, the charged oil droplets are pushed towards the oppositely charged electrode (in this case the positive electrode i.e. the anode). Also, the unidirectional electric field facilitates droplet-ion collisions helping in coagulation (Matteson et al., 1995; Mollah et al., 2001). On the other hand, in a.c. electric fields, there is no net electrophoretic force acting on the droplets.

Mathematical Model for Prediction of Oil Removal Rates.

The set of coupled equations was solved numerically using the fourth-order Runge-Kutta scheme, and the C_n values obtained from the model were fitted to the experimental data to obtain the value of single lumped parameter of the model, k_6. The fitting was started from a time $t > 0$. This was done to account for the induction time of ions (T_{ind}). Before this time there is no release of ions from the electrodes, and practically no oil is removed. Figures 2 shows a comparison between the values predicted by Matteson's model, and the model developed in this work. Clearly, the model developed in this work is predicting the experimental data to a higher accuracy. The correlation coefficient (r^2), induction time (T_{ind}), and the lumped parameter (k_6) values for various experimental runs are tabulated in table 4. The r^2 values ranged between 0.87 and 0.99, except for the run 7 with 15 V HWR a.c. across APE3, for which r^2 was only 0.14. From these high correlation-coefficient values it can be concluded, that the model developed for predicting the oil removal rate is in a good agreement with most of the experiments.

CONCLUSIONS

Various parameters influencing the electrocoagulator performance were analyzed, and a set of elecctrocoagulator's configuration and optimum working conditions for minimum SEC were identified. Electrode passivation was shown to greatly deteriorate the electrocoagulator performance (in case of APE), posing serious design problems. Also, APE electrode passivation and pitting were shown to reduce the electrode life to 6-7 runs. Electrode passivation and pitting effects were not present in the case of SSPE, providing a good alternative for APE. The 95 % oil removal time decreased with increasing voltage for both HWR a.c. and a.c. supplies. But, the SEC was shown to have an inverse relation with voltage for HWR a.c., and a direct relation with a.c. supply. A Pulse d.c. supply of 10 V, and frequencies of 50 Hz and 200 Hz, was used with SSPE. Although no appreciable change in 95 % removal times was observed on using pulse d.c. in place of HWR a.c., SEC decreased by a factor of three. Pulse d.c. did not show any significant variation on changing the 'pulse repetition frequency'. The most important conclusion from the experimental work was that huge savings can be made by using HWR a.c. supply instead of a.c. supply. The SEC reduces to 0.003 KW-Hr / liter (minimum obtained value for HWR a.c. supply) from 0.054 KW-Hr / liter (minimum obtained value for a.c. supply), a reduction of 94.44 %. This minimum value was achieved with a 5 V HWR a.c. supply across SSPE.

The mathematical model developed for predicting oil removal rates in the electrocoalescer was found to be in a good agreement with the experimental data reported in literature (Matteson et al., 1995) and those collected during the present investigation. The lumped parameter, k_6, did not show significant variation with applied voltage in the range of 5-15 V with SSPE. The experimental values were predicted to a good accuracy using an average value of k_6 for the applied voltage range of 5-15 V across SSPE. Electrode passivation, electrode material, and electrode area showed a strong effect on k_6 values, in agreement with their strong effect on oil removal rates. The sensitivity of k_6 values on the induction time, T_{ind} and the experimental difficulties in accurately measuring the T_{ind} values necessitate theoretical determination of induction time.

Table 1. 95 % - removal times and SEC for all the experimental runs

Electrode set	Run #	Operating Conditions	SEC (kW-Hr / Litre)	95 %-removal time (minutes), unless specified
APE1	3	40 V, a.c., with floating grease layer	-	45 (for 45 % removal)
	4	40 V, a.c., with skimming of the scum	-	130
	5	40 V, a.c.	-	148
APE2	1	40 V, a.c.	0.07	38
	2	10 V, HWR a.c.	0.018	80
	3	10 V, a.c.	0.171	660
	4	15 V, HWR a.c.	-	80
	5	25 V, HWR a.c.	-	58
	6	30 V, HWR a.c.	0.03	34
APE3	1	25 V, HWR a.c.	0.032	23
	2	20 V, HWR a.c.	0.023	39
	3	20 V, a.c.	0.074	200
	4	30 V,a.c.	0.054	81
	7	15 V, HWR a.c.	-	95 (for 20 % removal)
SSPE1	1	15 V, HWR a.c.	0.014	25
	2	25 V, HWR a.c.	0.03	33
	3	15 V, a.c.	-	540 (for 0 % removal)
	4	15 V, HWR a.c.	0.016	28
	5	15 V, HWR a.c., w/o innermost electrode	0.019	35
	6	15 V, HWR a.c.w/o outermost electrode	0.012	68
	7	10 V, HWR a.c.	0.011	57
	8	5 V, HWR a.c.	0.003	30
	9	10 V, pulse d.c., 50 Hz	0.004	28
	10	10 V, pulse d.c., 200 Hz	0.004	27
	11	10 V, pulse d.c., 150 Hz	0.0035	47
	12	10 V, HWR a.c.	0.008	130

Table 2. Current density and 95 % removal times for runs with different APE sets at 25V HWR a.c.

APE set	run #	Current density (Am^{-2})	95 %removal time (min)
2	5	99.5-66.3	59
3	1	207.3-249	23
3	5	49.8-33.2	65

Table 3. Replacement period of APE

APE set	# of runs	Replacement reason
1	5	Pitting
2	6	Pitting
3	7	Passivation

Table 4. Induction Time, T_{ind}, Correlation Coefficient, r^2, and lumped parameterk_6 values for various experimental runs

HWR a.c. across SSPE						
Voltage (V)	**# Electrodes**	**Electrode Area (cm^2)**	**Coagualtion type**	**T_{ind} (s)**	**r^2**	**k_6**
15	3	602.88	Shear + Brownian	600	0.95	7.08E-20
		602.88	Brownian only	600	0.84	3.59E-19
	2 (w/o outermost)	188.4	Shear + Brownian	600	0.9	8.45E-21
	2 (w/o innermost)	414.48	Shear + Brownian	600	0.97	3.28E-20
10		602.88	Shear + Brownian	600	0.91	3.27E-20
		602.88	Brownian only	600	0.83	8.74E-20
	STARTTIME					
5	1500	602.88	Shear + Brownian	1500	0.87	2.70E-20
	1800	602.88	Shear + Brownian	1800	0.93	4.30E-20
		602.88	Brownian only	1800	0.87	1.02E-19
	2000	602.88	Shear + Brownian	2000	0.96	6.24E-20
25	**Matteson's exptl data**	246	Shear + Brownian	-	0.99	1.89E-21
			Brownian only	-	0.99	6.13E-21
HWR a.c. across APE						
	Electrode set					
30	APE2, Run 6	602.88	Shear + Brownian	600	0.97	1.43E-20
25	APE2, Run 5	602.88	Shear + Brownian	700	0.96	2.43E-21
	APE3, Run 1	602.88	Shear + Brownian	400	0.97	4.57E-20
	APE3, Run 5	602.88	Shear + Brownian	1700	0.87	3.82E-21
20	APE3, Run 2	602.88	Shear + Brownian	1000	0.87	1.28E-20
15	APE2, Run 4	602.88	Shear + Brownian	1000	0.99	3.84E-21
	APE3, Run 7	602.88	Shear + Brownian	800	0.14	2.96E-29
10	APE2, Run 2	602.88	Shear + Brownian	1800	0.98	8.15E-21

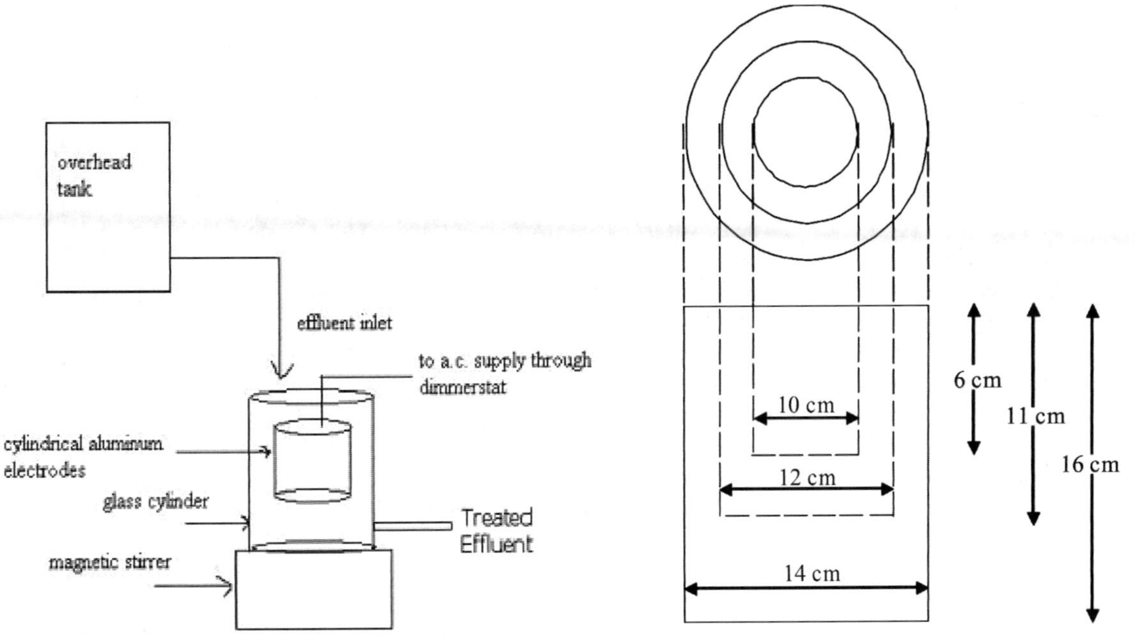

Figure 1: (a) Experimental apparatus and (b) concentric cylindrical electrodes assembly

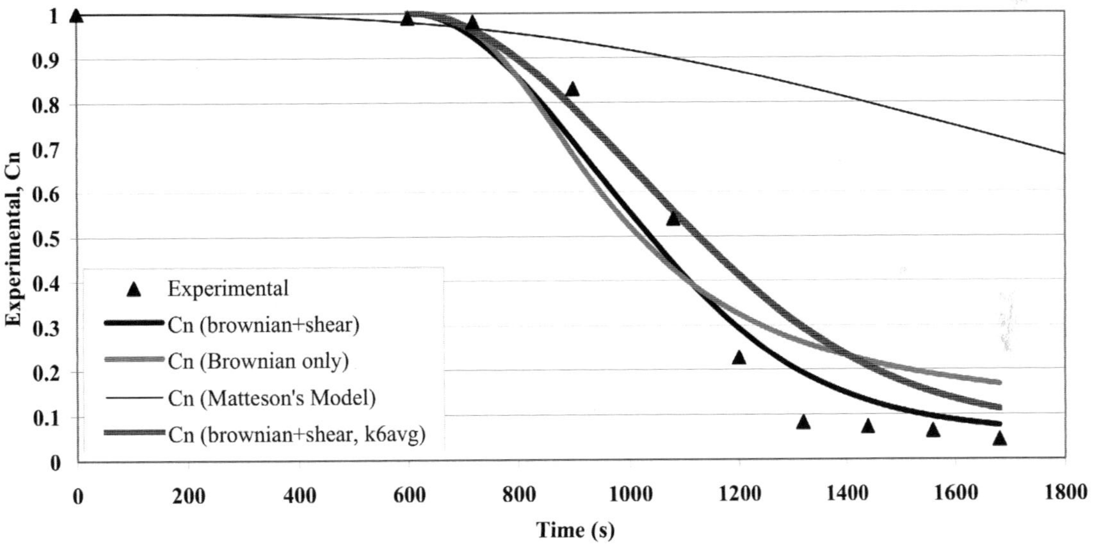

Figure 2: Comparison between the experimental and the predicted oil-removal rate at 15 V HWR a.c. across SSPE.

References

Amosov, V.V., Zil'brtman, A.G., "Experience in local treatment of wastewaters from petrochemical production", Chemistry and Technology of Fuels and Oils, **12,** 11- 12 (1976) 850-852.

Baklan, VY, Kolesnikova, IP, "Influence of electrode material on the electrocoagulation", *Journal of Aersol Science,* **27** (1986), Supplement 1: S209-S210.

Chen, X., Chen, G., and Yue, P.L., "*Investigation on the electrolysis voltage of electrocoagulation"*, Chemical Engineering Science **57** (2000) 2449 – 2455.

Do, J.-S., Chen, M.-L., "Decolourization of dye-containing solutions by electrocoagulation", Journal of Applied Electrochemistry, **24**, 8 (1994) 785-790.

Eow, S.J., Ghadiri, M., Sharif A.O., and Williams, T.J., "Electrostatic enhancement of coalescence of water droplets in oil: a review of the current understanding", Chemical Engineering Journal **84** (2001), 173–192.

Eow, S.J., Ghadiri, M., "*Electrocoalesce-separators for the separation of aqueous drops from a flowing dielectric viscous liquid"*, Separation and Purification Technology **29** (2002) 63–77.

Friedlander, S.K., "*Smoke, Dust and Haze*", John Wiley, New York (1977).

Holt, P., Barton, G., and Mitchell, C., "Electrocoagulation as a wastewater treatment", The Third Annual Australian Environmental Engineering Research Event. 23-26 November Castlemaine, Victoria 1999.

Jiang, J. Q., Graham, N., Andre, C., Kelsall, G.H. ,Brandon, N., "Laboratory study of electro-coagulation–flotation for water treatment", Water Research **36** (2002) 4064–4078.

Mameri et al., "Deflouridation of Sahara water by small plant electrocoagulation using bipolar aluminum electrodes", Separation and Purification Technology 24 (2001) 113-119.

Mameri, N., Yeddou, A.R. et al, "Defluoridation of septentrional Sahara water of North Africa by electrocoagulation process using bipolar aluminium electrodes", Water Research, **32**, 5 (1988) 1604-1612.

Matteson, M.J., Dobson, R.L., Glenn, R.W., Jr., Kukunoor, N.S., Waits, W.H. III, Clayfield, E.J., "Electrocoagulation and separation of aqueous suspensions of ultrafine particles", Colloids and Surfaces A: Physicochemical and Engineering Aspects **104** (1995) 101-109.

Mollah, M., Y., Schennach, R., Parga, J., R., Coke, D., L., "Electrocoagulation (EC)-Science and Applications". Journal of Hazardous Materials **B84** (2001) 29-41.

Nikolaev, N.V., Kozlovskii, A. S., Utkin, I. I., "Treating natural waters in small water systems by filtration with electrocoagulation", Soviet Journal of Water Chemistry and Technology, **4,** 3 (1982) 244-247.

Novikova, S.P., Shkorbatova, T.L., *et al.*, "Purification of effluents from the production of synthetic detergents by electrocoagulation", *Soviet Journal of Water Chemistry and Technology,* **4,** 4 (1982) 353-357.

Osipenko, V.D., Pogorelyi, P.I., "Electrocoagulation neutralization of Chromium Containing effluent", Metallurgis, **21,** 9-10 (1977) 44-45.

Probstein, R.F., "Physicochemical Hydrodynamics- An Introduction, Second Edition", John Wiley, New York (1994) P. 250, 251.

Qizheng, Ye, Jin Li, Haiyan Yang," A displaced dipole model for spheres in a non-uniform field", Journal of Electrostatics **56** (2002) 1–18.

Robinson, V., "*Electroflocculation in the Treatment of Polluted Water*", Electropure Australia Ltd.

Stoy, R.D., "Force on two touching dielectric spheres in a parallel field", Journal of Electrostatics **35** (1995) 297-308.

Vik, EA, Carlson, DA, Eikun, AS and Gjessing, ET, "Electrocoagulation of potable water", Water Research **18,** 11 (1984) 1355-1360.

Xiong Ya et al., "Treatment of dye wastewater containing acid orange II using a cell with three- phase three-dimensional electrode", Water Research. **35**, 17 (2001) 4226–4230.

Zolotukhin, I.A., (1989), "A pilot-scale system for the treatment of mine water by electrocoagulationflotation", Soviet Journal of Water Chemistry and Technology, **11**, 2 (1989) 147-151.

2b Controlled Particle Deposition by Design of An Electrochemical Adsorption Cell

Ehsan Bakhshi, R&D Center of PIDMCO ,National Petrochemical Company, Tehran, Iran and Fatemeh Abniki, Process Engineering Dept, National Petrochemical Company(ghadir group), No.38, Street 6, Dolat Abad, Shahrake Vahdat, Tehran, Iran.

Deposition of colloidal particles substrate is encountered in a variety of naturally occurring processes such as particulate fouling of heat exchangers, thrombus formation in vascular prostheses and other artificial organs and aerosols entering the lungs. Deep bed filtration of waste waters is perhaps the most prominent example of importance to industry.

Design of an electrochemical adsorption cell utilizing reticulated vitreous carbon (R.V.C.) as the working electrode ,stainless steel as the counter electrode and a cellulose acetate membrane ,for separating the anodic and cathodic, is described for the separation of polystyrene latex colloidal particles from a KCI solution.

The experimental results are compared with the Derjaguin-Landau-Verwey-Ovdrbeek (D.L.V.O.) prediction for the occurrence of favorable and unfavorable deposition conditions.

In the theoretical section of this work it will be assumed that the main energies contributing adsorption interaction between the surfaces will be due to electrical double layer to Van der Waals attraction . The some of these energies provides the total interaction energy between surfaces and this constitutes the underlying assumption for the classical D.L.V.O. theory of colloidal stability.

The polystyrene latex particles used in this study were 5.4 μm and hence effect of diffusion becomes negligible. For reasonably low flow rates, Re was less than l, inertial impaction is usually assumed negligible compared to the bulk flow. The highest particle Reynolds number employed in this study was 2800 and hence the contribution due to inertial impaction can be ignored . Therefore only the effect of gravity and electrostatic forces need to be considered in any theoretical analysis of the experimental results. Once the particles are in the vicinity of the collector. In order to observe solely the effect of these forces the effect of hydrodynamic interactions has to be kept at a minimum. Therefore experiments carried out were directed mainly at finding out the dividing line between hydro dynamically-controlled and electro statically-controlled processes.

In this way , for the present experimental conditions the main forces contributing to deposition process are the surface forces and hydrodynamics . At the highest flow rate of 34 cubic meter/min the colloidal forces were shown to be negligible compared to the hydrodynamics . Indeed the effect of surface at the KCl concentration of 0.001 mol/cubic dm apparently only became significant at the flow rate.

At this flow rate and KCl concentration of 0.001 mol/cubic dm experimental results show Clear evidence of conditions governing the of favorable and unfavorable deposition conditions . The variation of collector potential under unfavorable deposition condition does not produce large changes in the deposition rate whereas variation collector potential under favorable condition produce significant change in the deposition state.

2c Separation of p-Cresol from the Cresol Isomers by Stripping Crystallization

Lie-Ding Shiau, Department of Chemical and Materials Engineering, Chang Gung University, 259 Wen-Hwa 1st Road, Taoyuan, 333, Taiwan and Chih-Hao Huang, Chemical and Materials Engineering, Chang Gung University, 259 Wen-Hwa 1st Road, Taoyuan, 333, Taiwan.

Stripping crystallization (SC) is a new separation technique which combines distillation and crystallization. Basically, SC is operated at a triple-point condition, in which the liquid is simultaneously vaporized and crystallized due to the three-phase equilibrium. In essence, the process is continued until the liquid phase is completely eliminated and only the pure solid crystals remain in the feed. Thus, crystal washing is not required since only pure solid crystals remain in the feed and no impurities are adhered on the crystal surfaces at the end of SC. In the previous research, SC has been successfully applied to separate p-xylene from the mixed xylenes by Shiau et al. [1-3].

SC is applied in this study to separate the p-cresol/m-cresol mixture. Due to the close boiling points, it is very difficult to separate them by conventional distillation. The SC experiments show that p-cresol can be purified from 90% to 96%+ ,from 95% to 98%+ , from 97% to 98%+ , or from 98% to 99%+ in the p-cresol-enriched mixture. The experimental results are consistent with the simulation results. The recovery rate of p-cresol in the experiments can be greatly improved by adjusting the operating temperature and pressure during the SC process.

[1] Shiau, L. D.; Wen, C. C.; Lin, B. S., 2005, "Separation and purification of p-xylene from the mixture of m-xylene and p-xylene by distillative freezing", Ind. Eng. Chem. Res. , 44(7), p.2258-2265

[2] Shiau, L. D.; Wen, C. C.; Lin, B. S., 2006, "Application of distillative freezing in the separation of o-xylene and p-xylene", A.I.Ch.E. Journal, 52(5), p.1962-1967.

[3] Shiau, L. D.; Wen, C. C.; Lin, B. S., 2008, "Separation of p-xylene from the multicomponent xylene system by stripping crystallization", A.I.Ch.E. Journal, 54(1), p.337-342.

29a Horizontal Spinning Filter Discs for Solids/Liquid Separation

Robert Tarves, Mornach Environmental, Blackwood, NJ 08012

horizontal Spinning filter discs separate solids and fluids having different viscosities while requiring small footprint and low energy

29b Problems Arising out of Impurities in Steam Turbines

Fatemeh Abniki, Process Engineering Dept, National Petrochemical Company(ghadir group), No.38, Street 6, Dolat Abad, Shahrake Vahdat, Tehran, Iran, Ehsan Bakhshi, R&D Center of PIDMCO , National Petrochemical Company, Tehran, Iran, and Majid Mosalla, Material Science Dept., Shiraz University, Shiraz, Iran.

A long time research has been run on steam turbines for 22 months.In this way, many points found. Different impurities are transported into the steam turbine from the Preboiler cycle in the feedwater and the boiler as a total (mechanical plus

vaporous) carryover . Their major sources include condenser leak , make up water , and improperly operated condense polishers . The turbine environment is controlled through a control of impurity ingress and various feed water , boiler water , and steam chemistry limits . The corrosiveness of the steam turbine environment is caused by one or more of following :

1) Concentration of impurities from low part per billion levels in steam to percent levels on surfaces and the formation of concentrated aqueous solutions (concentration by deposition or evaporation of moisture)

2) Insufficient pH control (in both acid and alkaline regions)

3) High-velocity , high-turbulence , and low-pH moisture

The steam impurities that are of most concern include chlorides , sulfates , fluorides , carbonates , hydroxides , organic and inorganic acids , oxygen , and CO_2 .Their behavior in turbine steam and deposits is well documented .

There are strong synergistic effects and interactions with metal oxides .

In addition to corrosion during operation , turbines can corrode during :

1-Manufacture (machining fluids and lubricants)

2- Storage (airborne impurities and preservatives)

3- Erection (airborne impurities ,preservatives,and cleaning fluids)

4-Nondestructive testing (cleaning and testing fluids)

5- Lay up (deposits plus wet air)

Many of the above substances may contain high concentrations of sulfur and chlorine , which could form acids upon decomposition . Decomposition of typical organic for example , carbon tetra-chloride (CCl_4) , occurs at about 150 C . there fore , the composition of all of the above substances should be controlled and most of them should be removed before operation .

Web Page: www.nipc.net

29c Solids/Liquid Separation Challenges in Gas to Liquid Processes

Seyi A. Odueyungbo, Applied Research and Catalysis, Process R&D, Chevron Energy Technology Company, 100 Chevron Way, 10-1216, Richmond, CA 94802

Gas to liquid processes feature fischer-Tropsch reactors which transform syngas, a mixture of CO and H2, into hydrocarbon molecules. The reactor platform ranges from fixed bed to various forms of slurry reactors; the slurry bubble reactors (SBR) is used extensively. For GTL processes using slurry bubble column reactors, the finely graded iron or cobalt based catalyst is used. Solids liquid separation units-filters, settlers, hydrocyclone, etc.,- are used to enable the recovery of the Fischer tropsch products and facilitate catalyst recycle. The solids/liquid separation challenges of a typical gas to liquid process will be discussed.

AUTHOR INDEX

AUTHOR INDEX

American Institute of Chemical Engineers
3 Park Avenue
New York, New York 10016-5991

CURRAN ASSOCIATES INC
proceedings
.com